Technical, Commercial and Regulatory Challenges of QoS

The Morgan Kaufmann Series in Networking
Series Editor, David Clark, M.I.T.

For further information on these books and for a list of forthcoming titles, please visit our Web site at http://www.mkp.com.

Technical, Commercial and Regulatory Challenges of QoS

An Internet Service Model Perspective

XiPeng Xiao

AMSTERDAM • BOSTON • HEIDELBERG • LONDON
NEW YORK • OXFORD • PARIS • SAN DIEGO
SAN FRANCISCO • SINGAPORE • SYDNEY • TOKYO

Morgan Kaufmann Publishers is an Imprint of Elsevier

Morgan Kaufmann Publishers is an imprint of Elsevier.
30 Corporate Drive, Suite 400, Burlington, MA 01803, USA

This book is printed on acid-free paper.

British Library Cataloguing in Publication Data
A catalogue record for this book is available from the British Library

Library of Congress Cataloging-in-Publication Data
Xiao, XiPeng.
  Technical, commercial, and regulatory challenges of QoS / XiPeng Xiao.
    p. cm. — (The Morgan Kaufmann series in networking)
  Includes bibliographical references and index.
  ISBN 978-0-12-373693-2 (alk. paper)
  1. Computer networks—Quality control. 2. Computer networks—
Management. 3. Internet service providers. 4. Internet—Government policy. I. Title.
  TK5105.5956.X53 2008
  004.67'8—dc22

                                                                    2008026434

ISBN: 978-0-12-373693-2
ISSN: 1875-9351

For information on all Morgan Kaufmann publications,
visit our Web site at www.mkp.com or www.books.elsevier.com

Typeset by Charon Tec Ltd., A Macmillan Company. (www.macmillansolutions.com)

Printed and bound by CPI Group (UK) Ltd, Croydon, CR0 4YY

Transferred to Digital Print 2011

Contents

v

PART 3 THE NEXT STEP

List of Endorsements

The topic of QoS is all too often treated as a purely technical issue. In this refreshingly novel and comprehensive book, XiPeng Xiao draws on his own first-hand experience and that of other experts to put the technical issues in their correct commercial and regulatory context. This book is likely to make QoS much more understandable and relevant to a broad audience than it has been to date.
–Bruce Davie, Fellow, Cisco Systems

An admirable effort towards clarifying some of the key issues of Internet QoS.
–Daniel Awduche, Fellow, Verizon Business

I highly recommend this book filled with both technical and business insight.
–Zhiwei Yang, former CTO, China Netcom

Comprehensive and insightful discussion on QoS.
–Roger Wenner, IP Architecture, Technical Engineering Center, Deutsche Telekom

Finally, a QoS book that reflects network reality.
–Waqar Khan, Chief Architect, Qwest Communicating Inc.

It is really a wonderful piece of work. By providing many data network practical evidences, the author clearly explained the pros and cons in execution of net neutrality. This is the most comprehensive book that I have ever read on the net neutrality with a full taxonomy of implications related to users, OTT providers, ISPs and ASPs.
–Zhisheng Chen, Distinguished Member of Technical Staff, Sprint Nextel

The discussion on business model is helpful for the traditionally technical QoS subject.
–Alan Hannan, VP Engineering, Internap

An excellent piece of work from XiPeng, as usual.
–Dave Cooper, VP Network Architecture, Global Crossing

It is an excellent book on the business and technical challenges of QoS.
–Bill St. Arnaud, Senior Director, Advanced Networks, CANARIE

This book meets a real serious need in the QoS literature.
–Andrew Odlyzko, Director, Digital Technology Center, University of Minnesota, Interim Director, Minnesota Supercomputing Institute

This book fills a hole in existing QoS literature.
–Jennifer Rexford, Professor of Computer Science, Princeton University

The book is quite unique and impressive. It is quite readable and answers many questions that an engineer, a manager, a student, and an instructor may have.
–Lionel M. Ni, IEEE Fellow; Chair, Department of Computer Science & Engineering, Hong Kong University of Science and Technology

Comprehensive industrial view on QoS for academic researchers.
-Andrzej Jajszczyk, IEEE Fellow; Professor, AGH University of Science and Technology Poland; Former Editor-in-Chief, *IEEE Communications Magazine*

It is very well written and I can say we have the exactly same view.
-Dave Wang, President, WANDL (Wide Area Network Design Laboratory)

I highly recommend this book for its pragmatic analysis as well as its technical and "big picture" content.
-Arman Maghbouleh, President, Cariden

Preface

Today, the increasing popularity of mobile phones and VoIP generates a large impact on the revenue of traditional telecom service providers. To maintain their subscriber base and average revenue per user, telecom service providers are eager to offer premium services such as IP TV, online gaming, etc. It is assumed that these services will create a large demand for IP QoS. At the same time, there is a trend to use the Internet as the common carrier for all kinds of services, instead of having many special-purpose networks. It is assumed that this will bring the QoS requirement to the Internet. However, after so many years of research, development, and claimed deployment, QoS is still something of the future in the Internet. Among other issues, the service quality of the Internet can still be unpredictable. What makes QoS so elusive? What is missing? What needs to be done to bring QoS to reality?

The current Net Neutrality debate further complicates the matters on QoS. Since the idea of QoS was formed, it has always been taken for granted that if carriers can provide QoS, they can charge users for it. The Net Neutrality debate cast doubt on this belief for the first time. A Net Neutrality legislation can dramatically change the QoS landscape overnight. What is Net Neutrality? How will the debate shape the evolution of QoS? Why should a common person even care?

In this book, we will discuss the technical as well as commercial and regulatory challenges of QoS, and propose a model to overcome these challenges. We will first define what QoS is, and then discuss:

- What are the QoS requirements of common applications?
- What can the Internet offer in terms of QoS today?

This lets us see the performance gap and thus provides a base for subsequent discussions.

We then review the contemporary QoS wisdom, and discuss its commercial, regulatory, and technical challenges. Some of the important topics include:

- The commercial challenges of the traditional QoS wisdom, regarding:
  - Who should pay for QoS, business or consumers, senders or receivers?
  - Why does this matter?
  - What kind of assurance should carriers provide with QoS?
  - Will the attempt to sell QoS increase customer churn, because it is considered as evidence of poor quality for the basic service?

- The regulatory challenges of the traditional QoS wisdom, regarding:
  - What is Net Neutrality and how does it relate to QoS?
  - Will carriers be allowed to charge for QoS?
  - What is the impact of the uncertain government regulation on QoS?

- The technical challenges of the traditional QoS wisdom, regarding:
  - What cooperation is needed among carriers in order to provide QoS? Can they be done at an acceptable cost?
  - Will various QoS mechanisms introduce too much complexity into the network to hurt network reliability and reduce QoS rather than improve QoS?
  - What are the technical challenges to differentiating a higher CoS to a point that the end users can perceive the difference from Best Effort (so that the end users will be willing to buy the higher CoS)?

We will then propose how to improve the current QoS business model to overcome the commercial, regulatory, and technical challenges. On the commercial side, this involves a change to the QoS pricing scheme. On the technical side, this involves comprehensive consideration of all the options available, increased emphasis on certain mechanisms and deemphasis of some other mechanisms. We will go to great lengths to explain why the proposed pricing scheme is better for the industry and for the users, and back it up with a large amount of historic evidences. These evidences include revenue and usage statistics in the postal industry's 200-year history and in the telephony industry's 100-year history. These statistics establish the evolution trend of pricing schemes for communication services. We believe that our rationale for the proposed model becomes clear in light of the historic trend.

Next, we will present two case studies on real-world QoS deployment. One is about Internet2 (http://www.internet2.edu/), the next generation Internet test bed in the United States. This case study is written by Ben Teitelbaum and Stanislav Shalunov, formerly of Internet2 and now of Bit Torrent (http://www.bittorrent.com/). The other is about Internap, one of the few network service providers in the United States that successfully commercialized QoS (http://www.internap.com/). This case study is written by Ricky Duman of Internap. Because these case studies are written by the network operators who did the actual deployment, the readers can hear directly from the horse's mouth about QoS deployment issues in real-world networks, and the lessons they learned. We will also discuss QoS issues in wireless networks. That chapter is written by Dr. Vishal Sharma of Metanoia Inc. (http://www.metanoia-inc.com/), a well-known technology consulting firm in Silicon Valley. The contributions of Ben Teitelbaum, Stanislav Shalunov, Ricky Duman, Vishal Sharma, and Abhay Karandikar are gratefully acknowledged. We draw our conclusions at the end.

Throughout this book, there are a number of advanced technical issues that are discussed but are not fully resolved. These are good topics for further research.

Because this is the first book that covers all three important aspects of QoS—technical, commercial, and regulatory—and each aspect has a broad range of topics, we recognize that it is possible that over time, some of our opinions may turn out to be revisable. But we believe that this won't hurt the main purpose of this book, which is to help people see the big picture of QoS, think critically about QoS, and form their own opinion on QoS. With this recognition, we are eager to hear back from the readers. A web site has been set up at http://groups.google.com/group/qos-challenges for discussion. You can present your view points for other people to see.

AUDIENCE

In this book, we will discuss all three major aspects of QoS—technical, commercial, and regulatory—and how they interact with each other. We will first examine the status quo of QoS to show that the contemporary QoS wisdom has not been able to make QoS a reality for the Internet. We will then provide our explanation for this outcome by discussing the technical, commercial, and regulatory challenges. We will then propose a revision to the QoS model; discuss how it can overcome the commercial, regulatory, and technical challenges; and explain why we believe it is better for the industry.

We believe this book has value for the following audiences:

1. For people who are interested in understanding QoS technology, this book is a one-stop place for various flavors of technical QoS solutions, their pros and cons, the major overlooked factors in the current Diffserv/traffic management-centric solution, and the key trade-offs that must be made for a technical solution to be practical. The description is relatively high level so that most people can understand it without difficulty. For people who are interested in knowing the details, the book provides pointers to other references. The case studies about Internet2 and Internap's QoS deployments enable the readers to see QoS deployment issues in real-world networks. People about to deploy QoS can benefit from the lessons they provided. Academic people may also be interested in a number of advanced technical issues that are discussed but are not fully resolved—these can be good topics for further research.

2. For people who are interested in understanding the commercial issues, this book provides a comprehensive discussion about the commercial challenges in selling QoS. The key issues include "What is the effect of Internet users' Free mentality?," "What QoS assurance should be provided to attract users to buy QoS, soft or hard?," "Whom should NSPs charge QoS to, business or consumers? Senders or receivers?," "Will charging for QoS be considered as a sign of poor service quality for the basic service? Will it trigger customer defection?," "What should the contractual settlement among NSPs be to facilitate interprovider QoS?," "How much QoS revenue can realistically be generated?" These discussions are particularly useful for people/companies who plan to invest in QoS, for example, either developing QoS features or deploying QoS mechanisms.

3. For people who are interested in understanding the regulatory issues and the Net Neutrality debate, this book provides a succinct summary of the

key issues, and the opinions of both the proponents and opponents on these issues. This saves the readers from having to spend the time to locate the information and follow the discussions. This would help the readers quickly form their own opinions on Net Neutrality.

ORGANIZATION

This book contains three parts. Part 1 discusses the current situation of Internet QoS, and points out that the contemporary QoS wisdom has not been able to make QoS a reality. Part 2 explains this outcome by discussing the commercial, regulatory, and technical challenges. Part 3 proposes a revised QoS pricing scheme and a technical solution, and discusses how they overcome or relieve the commercial, regulatory, and technical challenges.

Part 1 contains four chapters.

- Chapter 2 discusses what QoS means in this book, common applications' requirements on QoS, and the degree that the current Internet meets those requirements. The purpose of discussing the application requirements is to make the objectives of QoS clear. The purpose of discussing the degree that the current Internet meets those requirements is to clarify the gap between what is needed and what is available, so that we know what else may be needed to deliver QoS.

- Chapter 3 discusses the historic evolution of QoS solutions. The purpose is to provide the readers with some technical background and a historic view on various flavors of technical QoS solutions.

- Chapter 4 discusses the "contemporary QoS wisdom," including its business model and its technical solution. This is to provide a base for commercial, regulatory, and technical examination.

- Chapter 5 discusses the reality related to QoS, especially from a commercial perspective. This is to give us a sense of how well the traditional QoS wisdom works.

Part 2 contains four chapters.

- Chapter 6 discusses the commercial challenges of the conventional QoS business model.

- Chapter 7 discusses the regulatory challenges.

- Chapter 8 discusses the technical challenges.

- Chapter 9 summarizes the key points discussed in this part, and discusses the lessons that are learned.

The purpose of discussing the commercial, regulatory, and technical challenges is to expose the issues of the conventional QoS model. The purpose of discussing the lessons learned is to point out the direction for possible improvement of the QoS model.

Part 3 contains five chapters.

- Chapter 10 proposes a revised pricing scheme for QoS, and discusses how it overcomes or relieves the most difficult commercial and regulatory challenges. To help the readers see the rationale of this revision, we present a large amount of revenue and usage statistics in the postal industry's 200-year history and in the telephony industry's 100-year history. These statistics establish the evolutionary trend of pricing schemes for communication services. Our rationale for the proposed pricing scheme revision becomes clear in light of the historic trend.

- Chapter 11 discusses the revised technical solution and its benefits.

- Chapter 12 presents two real-world QoS deployments at Internet2 and Internap, and the lessons they learned.

- Chapter 13 discusses QoS in the wireless world. Because network resource is much more limited in the wireless world, QoS approaches are very different too. This is another effort to help the readers see the big picture.

- Chapter 14 concludes the book.

It is recommended that this book be read in its entirety, and in the order the chapters are presented. This allows the big picture to manifest in a way that is easier to understand.

Acknowledgements

Writing the acknowledgements is the enjoyable part of this long and laborious journey. This book would not be possible without the help of a number of friends and experts.

First, I would like to thank Dr. Andrew Odlyzko. His Internet economics theory enlightened me. His research work formed the foundation of Chapter 10. He also reviewed the manuscript, provided valuable feedback, and pointed me to other information and subject matter experts.

I would like to thank Zheng Wang and John Yu for the helpful discussions during the conception of this book. Zheng Wang and Angela Chiu reviewed the proposal. Les Cottrell, K. Claffy, and Al Morton provided valuable information on Internet performance statistics and application QoS requirements for Chapter 2. Louise Wasilewski and Mike Benjamin provided early feedback on Net Neutrality. Jon Aufderheide and Brook Bailey provided information on service provider peering and service provider perspective. Dave Wang and Yakov Rekhter provided feedback on the key messages of the book. Zhisheng Chen and Yong Xue took the time to read the whole manuscript and provided valuable feedback. Bill St. Arnaud reviewed the manuscript and provided detailed comments and valuable suggestions for improvement.

Ben Teitelbaum and Stanislav Shalunov wrote the Internet2 Case Study. Ricky Duman wrote the Internap Case Study. These formed Chapter 12. Vishal Sharma and Abhay Karandikar wrote Chapter 13, "QoS in Wireless Networks." Their generous contributions are greatly appreciated.

Rick Adams, Rachel Roumeliotis, and Greg Chalson at Elsevier managed to drag me to the finish line. Without them, this book would still be in writing. Rick was very involved in the development of the manuscript, and provided useful guidance and candid feedback.

Last but not the least, I would like to thank Amy Chen for taking care of our son Alan Xiao, thus allowing me to concentrate completely on the book during the crunching stage.

XiPeng Xiao

About the Author

XiPeng Xiao has a unique background in QoS. He did his Ph.D. thesis on QoS at Michigan State University. This gave him a strong theoretical background on QoS. The author has product management experience with multiple network equipment vendors in Silicon Valley and network operations experience with a global network service provider. This vendor experience let him know the implementation cost of various QoS mechanisms and the hidden caveats behind those mechanisms. The NSP experience let him understand the practical trade-off that network operators must make between the need for network control and the need for network simplicity. The author also participates in international standard organizations such as the Internet Engineering Task Force (IETF), the standard organization that drives the technology development of IP networking. This let him know what's going on in the industry in terms of QoS, and what other people are thinking and doing. But maybe most importantly, he is a technologist-turned-marketing person. At the network equipment vendors, he is responsible for managing product lines for market success. This forced him to look beyond technology and develop the sensitivity to business and regulatory issues. This unique background with research experience, vendor experience, provider experience, standard experience, and business experience enables him to see the big picture of QoS, which comprises of a technical aspect, a commercial aspect, and a regulatory aspect.

With this unique background, the author has made contribution to QoS before writing this book. In 1999, while deploying QoS and MPLS Traffic Engineering in Global Crossing's network, the author and his colleagues discovered a conflict between Diffserv and TCP. They proposed a solution which was eventually standardized as [RFC2873]. This book is his continuous effort to describe the QoS big picture. In 1999, he published "Internet QoS: A Big Picture" in *IEEE Networks Magazine*. According to Google Scholar (scholar.google.com), it is among the most quoted QoS articles. Over the years, the author published multiple RFCs and journal articles in the fields related to QoS. The author also made many presentations on QoS in many industrial forums and conferences.

Introduction

1

This chapter will discuss why we need a big picture of Internet Quality of Service (QoS) and will provide a high-level overview of how the Internet works for people who are not familiar with it.

THE BIG PICTURE

In November 2005, in an interview with *Business Week*, Edward E. Whitacre, Jr., Chairman and CEO of SBC (now AT&T), stated:

> *"Why should they be allowed to use my pipes? The Internet can't be free in that sense, because we and the cable companies have made an investment and for a Google or Yahoo! or Vonage or anybody to expect to use these pipes free is nuts." [Bizweek].*

This statement triggered a Net Neutrality debate between the Internet Content Providers (ICPs) (e.g., Google, Yahoo!, etc.) and Network Service Providers (NSPs) (e.g., AT&T, Comcast, etc.). The debate soon spread into the general public, the U.S. Senate, the House of Representatives, and the Federal Communications Commission (FCC). To date, controversial Net Neutrality legislation is still being discussed.

One of the important aspects of Net Neutrality is whether NSPs should be allowed to charge a QoS fee, and prioritize users' traffic accordingly. Many people in the IT/network/telecom industries actively participate in this debate. A large amount of online discussions have been generated.

From following the debates on Net Neutrality/QoS, we frequently see insights mixed with misconceptions, and facts mixed with personal beliefs. It is probably safe to say that the people who take the time to express their opinion on Net Neutrality have above-average knowledge on QoS. After all, they at least formed their own opinion. The fact that their opinions are so divergent indicates that after many years of research, standardization, implementation, and claimed deployment, QoS is still elusive and therefore intriguing.

1

The following is a list of topics that are vigorously debated in the Net Neutrality controversy. As a reader, if you agree to many of the following points, then this book can be useful for you, because it will provide the opposite views on these topics. This is not to say that the following points are all wrong, but to say that there is the other side of the story. After reading this book, you will get some new perspectives, and you may find that some of these points are not so agreeable anymore.

- The basic goal of QoS is to use Differentiated Service (Diffserv) and traffic management to create one or multiple Classes of Services (CoS's) that are better than Best Effort.
- NSPs need to invest a large amount of resources to provide QoS; they should be allowed to sell QoS to recoup some of their investment.
- If NSPs are not allowed to sell QoS, they will just provide raw bandwidth and will be reduced to "dumb pipers" like utility companies.
- QoS is a commercial and technical matter; government should stay out of it.
- Selling QoS will bring in additional revenue for the NSPs. The money will trickle down the value chain. Therefore, it's good for the whole industry.
- When different traffics are prioritized differently, users will be able to tell the difference between the different CoS's to pick the right CoS for their applications.
- Consumers and enterprises who want better service than today's Best-Effort service will buy a higher CoS if one is available.
- It doesn't matter whether business or consumers, senders or receivers, pay for QoS, as long as somebody pays for it.
- Traffic should be prioritized so that if there is congestion, high-priority traffic can be preferred. This can't be wrong.
- Relying on sufficient capacity to deliver QoS (i.e., over-provisioning) is a more expensive QoS approach compared to others.
- TCP is "greedy" (i.e., it will hunt for the largest possible sending rate). Therefore you can never have too much bandwidth because TCP will eat it up.
- Because data traffic is self-similar (i.e., still very bursty even after a large amount of aggregation), transient congestion is unavoidable.
- Network reliability/availability is important. But it is orthogonal to QoS. QoS and reliability can be tackled separately.
- QoS is not a reality today because the QoS mechanisms provided by network equipment vendors are not matured enough for the NSPs to deploy. When such mechanisms mature, NSPs will deploy them, and QoS will become reality.

From our perspective, the divergent QoS opinions indicate that people have a silo-ed view on QoS. That, in turn, indicates that existing QoS literature didn't present a QoS big picture.

First, there is a lack of coverage on the commercial and regulatory issues. QoS literature is too focused on the technical side. Since 1998, more than ten QoS books have been published [Ferguson][Armitage][Huston][Vegesna][Wang][Jha]

[Park][Szigeti][Alvarez][Soldani][Evans]. But none of them has coverage on the regulatory side and only a few [Ferguson][Huston] have scarce coverage on the commercial side.

The lack of coverage on the commercial side may be a little puzzling, given that there are a fair number of discussions on it. For example, the difficulty in selling QoS as an add-on to network connectivity has long been discussed among network operators. This is considered part of the "free mentality" problem of the Internet (for NSPs): After buying network connectivity, Internet users think that everything else on the Internet should be free. Also, since the inception of QoS, there is always this debate of "QoS vs. no QoS." (This name can be a little deceiving. The debate is really about "use mechanisms to provide QoS" vs. "use capacity to provide QoS".) This debate involves technical as well as commercial issues, that is, which approach is more feasible and which approach is more economic. Given these, the lack of commercial coverage in the QoS literature can be a little puzzling. We believe that this is partly because network operators are not very keen on contributing to the QoS literature. Furthermore, certain debates are done in a philosophical or even religious way, not in a logical way. For example, in many cases, each debating party simply takes its own assumptions for granted and does not bother to validate them. The basic argument became "I am right and you are wrong. I know it." There is little value to add that to the QoS literature.

The lack of coverage on the regulatory side is less of a surprise. After all, the Net Neutrality debate only started at the end of 2005 and the earliest Net Neutrality legislation was proposed in 2006. So it is no wonder that some people are shocked to hear that the government would even consider forbidding NSPs from selling QoS. "Why does the government have any business in QoS?," these people would think. The truth is, as the Internet became an integral part of ordinary people's daily life, traffic prioritization and the related pricing (i.e., QoS) can also affect ordinary people's daily lives. Because of its relevance to the mass people, QoS inevitably becomes a topic of interest to government regulators. If we further consider the phenomenon that some people think that introducing traffic prioritization would move the Internet from being neutral today to being unneutral tomorrow, and would deprive Internet users of the freedom to pick their favorite applications and application providers, then the government's involvement should not be a surprise. This is not to mention that the charging of QoS, or the lack of, involves an interest conflict between the ICPs, such as Google and MSN, and the NSPs, such as AT&T and Verizon, both with deep pockets and strong lobbying power. Therefore, the fact that some people are shocked does not say that government involvement is absurd. It just reflects a void of regulatory coverage in the QoS literature.

Second, even on the technical side, the technical big picture has not been clear. Technical QoS literature has been too focused on using Diffserv and traffic management to deal with traffic congestion. But many important questions are left unanswered. For example, how often will congestion happen? If congestion is too rare or too often, then Diffserv may not be useful. (If it is too often, something else is fundamentally wrong, and that needs to be fixed first.) Also, when congestion happens, is it better to handle it with Diffserv and traffic management, or is it better to find some other ways to relieve congestion? If the latter can be done most of the time, Diffserv/traffic management won't be useful either. Also, will the network have too much high-priority traffic to render Diffserv ineffective? If video becomes priority traffic, this would become true. The failure to ask and answer these questions indicates a lack of a technical big picture.

To a certain extent, we believe that this lack of a big picture is somewhat inevitable. Modern people are highly specialized. Few people have the opportunity to develop a comprehensive view on QoS. Today, the majority of QoS literature is authored by academic people. But academic people generally do not have the opportunity to develop a comprehensive industrial view. Economic and legal experts may have an industrial view. But they may not have the technical depth to tackle QoS head on. Their works are relatively focused on the economic and legal aspects of Internet services, and may be considered by normal QoS people as irrelevant to QoS. Industrial experts may be the most suitable people to present the big picture. But it is still not easy to find people with a strong background in all three aspects: technical, commercial, and regulatory. The people who have such expertise may not have the time to write either.

The effect of this lack of a big picture is inefficiency. We list a few examples below.

First, without a QoS big picture, it is difficult for people to communicate effectively. It is rare that people clearly list their assumptions when they express their opinions. They either don't realize that their arguments are based on certain assumptions, or assume that the other party accepts the same assumptions. Therefore, if both parties have a silo-ed view, and their silos are different (i.e., their assumptions are significantly different), then they can only wonder why the other party "just doesn't get it." This happens in many online Net Neutrality debates. This is also why some people are surprised to find that there are other people who would disagree with the bulleted points listed at the beginning of this chapter—those points seem like truisms to them.

Second, without a QoS big picture, people may come to a suboptimal decision on QoS. For example, on the technical side, if some network operators think that Diffserv/traffic management is the primary way to enable QoS, they may introduce a lot of Diffserv/traffic management mechanisms into their network. By the time they realize that Diffserv/traffic management alone cannot provide a satisfactory QoS solution, too much complexity may have crept into the network. On the commercial side, if a NSP fails to recognize that there are significant commercial challenges in selling QoS, it will likely fail to prepare a strategy to deal with those challenges. Consequently, its technical investment to enable QoS may not get good return. On the regulatory side, if some NSP people do not fully understand the argument of the Net Neutrality proponents, they may think that Net Neutrality has little validity. In an effort to get it over with, those people may attack Net Neutrality. That will have the exact opposite effect. The harder anybody tries to discredit Net Neutrality, the stronger its momentum will be. NSP's best strategy to avoid Net Neutrality legislation may be "do nothing"—at least in public.

Third, without a big picture, current QoS models overlooked several key factors. This reduces the commercial viability of QoS. That may be the reason why there are few commercial QoS success stories today. One of these factors is "user perceivability." When traffic prioritization/class of service is proposed, it is automatically assumed that their effect can be clearly perceived by the end users. "User perceivability" is critical because without it, users won't buy QoS (i.e., the higher CoS). The lack of effort to make sure that the effect of a QoS solution can be clearly felt by the end users reflects a lack of commercial consideration, and thus a lack of a big picture. Reliability is often another major overlooked factor. A significant portion of today's QoS problems are caused by human errors and equipment failures. QoS mechanisms can introduce additional complexity and reduce reliability. But existing QoS literature rarely considers the reliability impact of the proposed QoS mechanisms. This reflects a lack of a technical big picture.

Therefore, it is desirable to have a QoS big picture. This book is an effort towards that.

HIGH-LEVEL OVERVIEW OF HOW THE INTERNET WORKS

In this section, we give an overview of how the Internet works for people outside the network industry. This includes what the Internet looks like and how the major control components fit together.

At the highest level, the Internet consists of a number of Autonomous Systems (ASs) interconnected together, as shown in Figure 1-1.

Each AS is basically a NSP, e.g., AT&T or Deutsche Telekom. Each AS has its own customers. For example, many residential users are customers of AT&T or Verizon, while Google and Yahoo! are customers of Level3. Note that on the Internet, each customer is represented by its IP address. Via a protocol called external Border Gateway Protocol (eBGP), each AS tells other ASs who their customers are. Each AS also helps other ASs propagate their words. For example, AS2 in the

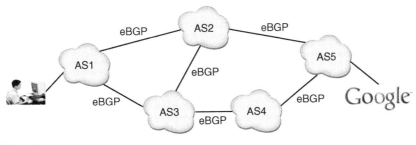

FIGURE 1-1

High-level view of the Internet

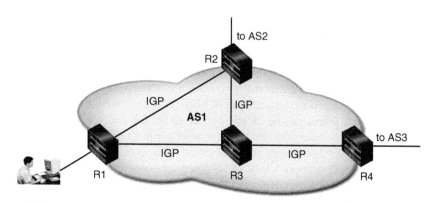

FIGURE 1-2

Network and control inside an AS

picture will tell AS1 that Google is a customer of AS5. This way, each AS will eventually learn which customer is in what AS. Because AS1 will hear from both AS2 and AS3 that Google is at AS5, AS1 also knows that it can get to Google through both AS2 and AS3. BGP also tells AS1 that to get to Google through AS2 involves two ASs, AS2 and AS5, while to get to AS5 through AS3 involves at least three ASs. All things being equal, AS1 will pick AS2 because the AS_Path is shorter. However, network operators at AS1 can override this decision. For example, if AS3 charge AS1 less money per unit of bandwidth, AS1 may decide to send all traffic to Google through AS3 instead. The use of a policy to override default routing protocol decision is called Policy-Based Routing.

Now let's zoom into each AS, using AS1 as an example, as shown in Figure 1-2.

For discussion purposes, let's assume that AS1 has four sites, each represented by a router, partially meshed together as showed in Figure 1-2. R1, R2, and R4 have connections to the outside world and have learned some routes to the external destinations. Inside AS1, these routers use the internal BGP (iBGP) to exchange reachability information. With iBGP, R1 will learn that to get to destinations in AS3, its next hop router should be R4 although it is not directly connected. How R1 can get to R4 inside AS1 is resolved by an Interior Gateway Protocol (IGP). The commonly used IGPs are Intermediate System-to-Intermediate System (IS-IS) [RFC1195] and Open Shortest Path First (OSPF) [RFC2328]. The reason both BGP and IGP are used inside an AS is that BGP is for managing external routes (i.e., external destinations) and IGP is for managing internal routes. Inside an AS, both types of routes are involved, therefore both protocols are needed.

BGP and IGP jointly determine the traffic distribution inside an AS. Sometimes the network operators of an AS may want to change the traffic distribution. For example, there may be too much traffic over the R1-R2 path for some reason. Therefore, the network operators want to offload some of the traffic from the R1-R2 path to the R1-R3-R2 path.

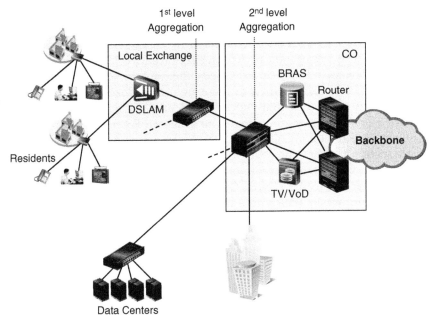

FIGURE 1-3

Metro network architecture

Usually, the network operators would start with IGP by making the direct R1-R2 path and the R1-R3-R2 path have equal IGP metric (i.e., length). This would cause traffic from R1 to R2 to be distributed on both paths. This practice is called IGP traffic engineering. But sometimes making multiple paths have equal cost can be difficult to do or can have undesirable side effects. This is why Multi-Protocol Label Switching (MPLS) traffic engineering is introduced. MPLS traffic engineering allows the use of multiple paths with different IGP metrics for load sharing. This kind of intra-domain traffic engineering is transparent to the outside world.

Sometimes, the network operators can also use BGP to change traffic distribution. For example, the reason there is too much traffic from R1 to R2 may be AS1 is using AS2 (connected via R2) to get to too many destinations. AS1 may decide to get to some of those destinations via AS3 instead by changing its policy. Naturally, the amount of traffic from R1 to R2 will be reduced. This kind of inter-domain traffic engineering will affect traffic distribution in other NSPs.

The above are concerned with the Internet backbone, i.e., the Wide Area Network (WAN), which covers multiple metropolitan areas. Now let's zoom into the Metro Area Network (MAN). Figure 1-3 provides a detailed metro network diagram. It is provided because not many people know what a metro network looks like.

In the metro, there are three groups of users, residential, enterprises, and data centers. The first two groups are content users while the third are the content providers. Residential users are connected via Digital Subscriber Line (DSL) or cable modem or Passive Optical Network (PON). The access devices, e.g., the DSL Access Multiplexer (DSLAMs), are connected by an aggregation network to the Central Office. The aggregation network usually has two levels of devices, with the first-level devices in the Local Exchange Office and the second-level devices in the Central Office. In each metro area, there are usually dozens of Local Exchange Offices and one or two Central Offices. Residential aggregation networks used to be based on ATM. They are migrating to Ethernet. Enterprises' access routers are connected to the Central Office via TDM links (e.g., T1/E1's) or Ethernet links. Similarly, servers in Data Centers are first aggregated by an Ethernet switch fabric, and connected via an Ethernet link to the Central Office. Note that although all the links connecting to the Central Office are showed as point-to-point links at Layer 2, they are usually transported over a ring-like optical network at Layer 1, with Ethernet mapped over SONET/SDH if necessary. Sometimes, direct fiber links may be used for Ethernet. This Layer 1 transport network is usually common for residential, enterprise, and data center customers. Also, to increase network availability, each device in the metro aggregation network may be connected to two higher-level aggregation devices, although this is not showed in the diagram. For example, the DSLAM may be connected to two first-level aggregation devices, and each first-level aggregation device may be connected to two second-level aggregation devices.

Inside the Central Offices, in addition to the second-level aggregation devices, there are also routers, Broadband Remote Access Servers (BRAS's), Video on Demand (VoD) servers, and possibly other servers. The BRAS's are the control devices for residential users' Internet access service. Other servers are for their own respective services. These servers are usually dual-homed to the routers. The routers are those shown in Figure 1-2. They are connected to routers in other metro areas.

Metro aggregation networks are usually Layer 2 networks. Therefore, there is no routing protocol involved. From a Layer 3 perspective, customer's access routers (i.e., the residential DSL modems or the enterprise CPE routers) are directly connected to the BRAS and the routers in the Central Office, respectively.

Up to this point, we have presented what the Internet looks like and how the major control protocols fit together to work. To recap, each domain (i.e., AS) has customers connected to it in some way, as shown in Figure 1-3. Each domain will announce its customers' IP address via BGP to other domains so that every domain knows what domains to go through to get to each destination. Inside each domain, IGP tells each router how to send packets to another router. Together, BGP and IGP determine the packet route between any two points. This way, communication between any two end points can ensue.

In relation to QoS, we would like to point out that the likely communication bottlenecks in the Internet are the links between the DSLAMs and the first-level aggregation devices, and the links between ASs. The former can become a bottleneck because NSPs assume that their residential users will rarely send/receive at maximum speed, or at least not at the same time. Therefore, they allow the sum of all customers' maximum access speed to be much higher than the DSLAM uplink's speed. Sometimes the ratio can be as high as 20. However, peer-to-peer file sharing is causing this assumption to be violated. Consequently, the DSLAM uplinks may become congested. The links between ASs can also become bottlenecks because upgrading them involves agreement between two different organizations. Sometimes that can take a long time. Consequently, traffic outgrows link speed and congestion happens. But, in general, network operators have a mental problem with chronic congestion in their network. So they will try to upgrade link speed to relieve that, no matter how tight the budget is.

Sometimes, the last-mile links, e.g., the DSL local loops or the enterprise access links, can also become bottlenecked. But today, most customers have the option to upgrade to a higher-speed link. Therefore, when congestion happens at the last mile, it is usually by the customer's own choice, mostly to save money.

If you have questions, please ask them at http://groups.google.com/group/qos-challenges.

The Status Quo

This part contains four chapters:

- Chapter 2 discusses what QoS means in this book, common applications' requirements on QoS, and the degree to which the current Internet meets those requirements.
- Chapter 3 discusses the historic evolution of QoS solutions.
- Chapter 4 discusses the current "mainstream QoS wisdom," including its business model for QoS and technical solution for QoS.
- Chapter 5 discusses the network reality related to QoS, especially from a commercial perspective.

The purpose of discussing the QoS requirements of the applications is to make the objectives of QoS clear. The purpose of discussing the degree to which the current Internet meets those requirements is to let us know the gap between what is needed and what is available, so that we know what else may be needed to deliver QoS. The purpose of discussing the historic evolution of QoS solutions is to provide some technical background. The purpose of discussing the current QoS business model and its technical solution is to provide a base for commercial, regulatory, and technical examination. The purpose of discussing the commercial reality related to QoS is to give us a sense of how well the traditional QoS wisdom works.

What Is QoS?

2

QoS can mean different things to different people. In this chapter, we first define what QoS means in the context of this book. We then discuss:

- What are the factors that determine the end users' QoS perception?
- What are the QoS requirements of common applications regarding these factors?
- What can common IP networks offer regarding these factors without active traffic management?

The first two topics will provide the objectives for QoS. After all, QoS itself is not the goal. The goal is to make sure that the end users' applications work satisfactorily. Therefore, we need to know the factors that determine the end users' QoS perception, and the common applications' requirements regarding these factors. The third topic establishes a baseline for QoS delivery. If we know what an IP network can offer under normal condition without applying any traffic management mechanisms, then we can compare it with the requirements, and decide what else is needed to provide QoS to end users.

QOS IS GOOD USER PERCEPTION

This book defines QoS from an end-user perspective. In this book, QoS means good network service quality for the end users. More specifically, if we say that a network has QoS or provides QoS, we mean that the network is able to meet the need of the end users' applications in a satisfactory way. With this definition, QoS has a broad scope. Anything that can affect end user's perception on network service quality, for example, reliability, security, routing policy, traffic engineering, all fall into the scope of QoS.

QoS, though, can be defined differently from other perspectives. One common definition defines QoS as using various traffic management mechanisms to create multiple Classes of Service (CoS) and differentiate them. People defining QoS this way acknowledge that reliability, security, routing policy, traffic engineering, etc. can all affect the end users' perception on network service quality. But they choose to take a "divide and conquer" approach, in which "QoS" people will focus on traffic management, and other people will take care of reliability, security, routing, traffic engineering, etc. Together, QoS and other mechanisms will work to provide the Quality of Experience (QoE) to the end users.

While there is nothing wrong with this perspective technically, we believe that defining QoS from an end-user perspective can be more productive. First, networks are built to meet the needs of end users. This calls for a user-centric perspective. End users care about network service quality, but they rarely distinguish between the terms QoS and QoE. Therefore, it is easier just to equate QoS to network service quality. Second, by including all factors that can affect end-user perception on network service quality into the scope of QoS, the interaction between these factors can be explicitly considered. For example, when adding certain traffic management mechanisms to enhance end-user experience, complexity will also be added to the network. This complexity may reduce the network reliability, which will affect end-user experience. Therefore, an explicit tradeoff must be made regarding how much traffic management to introduce.

This book also explicitly distinguishes between QoS and QoS mechanisms. QoS is the goal, whereas QoS mechanisms are the means. That is, QoS mechanisms are specific methods, such as traffic management, routing, and traffic engineering, to achieve QoS. It is not uncommon to hear somebody at an Internet Engineering Task Force (IETF) or North America Network Operators' Group (NANOG) meeting say that "There is no need for QoS; all you need is sufficient capacity." In this book's terminology, the sentence would be worded as "There is no need for QoS mechanisms; all you need to achieve QoS is sufficient capacity."

Another important point to note is, with the QoS definition in this book, a QoS solution has two main parts: the network part and the end-point part. The network part is concerned with what the networks can do for the applications. The end-point part is concerned with what the end points (for example, TCP stack, applications themselves) can do for the applications. This latter part is mainly concerned with the software application industry, not the network industry. This part of the QoS solution can be very diversified and application-dependent. It will not be the focus of this book. This book will focus on the network part because it is common to all applications.

WHAT FACTORS DETERMINE THE END USERS' QOS PERCEPTION?

Different people's perception of the same thing can be different, because perception inevitably involves some subjective factors. These subjective factors are usually difficult to quantify.[1] Fortunately, end users' perception of the performance of their network applications is largely determined by a number of factors such as delay, delay variation, and packet loss ratio that can be quantified. These factors are discussed below.

Delay

Simply put, end-to-end delay is the time needed for the destination application to get the packet generated by the source application. The scientific and precise definition of end-to-end delay can be a little complicated. Because the resulting difference is small, we simply use the simple and intuitive definition above. End-to-end delay consists of two parts: end-point application delay and network delay. Sometimes the delay that people refer to is just the network delay.

End-point delay is the delay introduced by the end-point applications. For example, packetization of a voice sample takes time and introduces the packetization delay. An application often introduces a jitter buffer to reduce jitter (that is, delay variation), and the jitter buffer will cause buffering delay.

Network delay is defined as the time the first bit of the packet is put on the wire at the source reference point to the time the last bit of the packet is received at the receiver reference point [RFC2679]. This is referred to by ITU-T as IP Packet Transfer Delay (IPTD), and sometimes explicitly as one-way delay. Network delay can be further divided into three parts:

- Transmission delay, that is, the time it takes to transmit a packet into the wire. This part can be significant for large packets over low-speed access links such as DSL or T1 lines. For example, a 1500-byte packet transmitted over a 380 Kbps DSL upstream link would take about 32 ms. Transmission delay is insignificant for high-speed links. For example, a 1500-byte packet transmitted over a 155 Mbps STM-1/OC-3 link would take only 0.08 ms.

- Packet processing delay, that is, the time it takes to process a packet at a network device, for example, queueing, table lookup, etc.

- Propagation delay, that is, the time it takes for the signal to travel over the distance. Because electromagnetic signal (for example, light) that carries the data does not propagate as fast in fiber or copper as in a vacuum, a rule of thumb is a propagation delay of 5 ms will be added every 1000 km of fiber route.

In summary:

$$End\ to\ end\ delay = End\text{-}point\ delay + Network\ delay$$

---

[1] But the International Telegraph Union's (ITU)'s [P.800] "Methods for Subjective Determination of Voice Quality" does provide a method. In P-800, an expert panel of listeners rated preselected voice samples of voice encoding and compression under controlled conditions. A Mean Opinion Score (MOS) can range from (1) bad to (5) excellent, and an MOS score of (4) is considered toll quality.

From an end-user perspective, delay determines the applications' responsiveness to users' actions, for example, how fast the listener can hear the talker's words, or how fast a web user can see some response when he/she clicks on a web link.

Different applications have different requirements on delay. This will be further discussed later in this chapter. Generally speaking, for real-time applications, end-to-end delay below 150 ms would ensure good service quality; an average delay above 400 ms is generally not acceptable for interactive applications.

While each packet has a delay value, it is generally not measured. To find out the delay performance of a network, specific measurement packets must be injected. When reporting the average delay or maximum delay, the length of the measurement period can make a big difference (for example, 1 minute vs. 5 minutes). Therefore, when reporting the delay performance of a network, the specific measurement method must also be described. One recommended measurement approach is to report delay metric as the arithmetic mean of multiple measurements over a specified period of time. Error packets and lost packets are excluded from the calculation. In [InterQoS], a group of network experts recommend that:

- Maximum evaluation interval = 5 minutes
- Mean packet separation = 200 ms
- The delay metric is reported to 1 ms accuracy

Delay Variation

Every packet has a delay (that is, IPTD) value. The IP Packet Delay Variation (IPDV) of a packet is the difference between its delay and the delay of a selected reference packet. For a flow, that is, a stream of packets belonging to the same application, the packet with the lowest IPTD is usually selected as the reference packet. In other words, because delay varies, delay variation serves to tell people the distribution of delay values. From this perspective, delay variation is ancillary to delay.

Delay variation can be caused by the following factors:

- Different packets can have different queueing delays at the same networking device.

- Different packets can have different processing delays at the same networking device. The difference will usually be small for modern networking devices that use hardware-based packet forwarding but can be big for legacy networking devices that use software-based packet forwarding with caching.

- Different packets may travel via different network paths and accumulate different queueing delays and propagation delay.

From an end-user perspective, delay variation determines the consistency of applications' responsiveness. When delay variation is small, applications' responsiveness, no matter how fast or slow, is consistent. This makes it easier for the end users to adapt to the communication situation at that moment. For example, when talking over a satellite link, the delay is usually long (for example, 400 ms). However, if delay variation is small, the talker can estimate how long it will take the listener to hear him or her, and determine whether the listener wants to talk or listen. The talker can then decide whether to continue talking or switch to listening. Consequently, although the delay is long, the conversation can go on. In contrast, if the delay variation is large, then after talking, the talker would not know whether the ensuing silence is caused by an unusually long delay, or is an indication that the listener wants to continue listening. The talker can only guess, and if the guess is wrong, the talker and listener will end up either interrupting each other or waiting for each other to talk. Either way, the communication will become difficult.

Different applications have different requirements on delay variation. This will be further discussed later in this chapter. Generally speaking, end-to-end delay variation below 50 ms would ensure good service quality even for delay variation-sensitive applications. Some applications don't have significant delay variation requirements.

Like delay, delay variation must be measured. [InterQoS] recommends the following measurement method:

- Maximum evaluation interval = 5 minutes
- Mean packet separation = 200 ms
- The delay variation metric is reported in ms, accurate to 1 ms and rounded up.
- One delay variation value, the 99th percentile value, is reported for each test period.

Packet Loss Ratio

Packet Loss Ratio (PLR) is the percentage of packet lost while traveling from the source to the destination.

Packet loss can be caused by the following factors:

- Certain links are congested, causing packet buffer overflow at the networking device.
- Some network device fails, causing packets inside the device to be dropped.
- Network reachability status changes (for example, routing protocol re-converges), causing certain packets to become "destination unreachable" and be dropped.

From an end-user perspective, packet loss ratio determines the presentation quality of the user applications, for example, sound quality of voice and picture quality of video.

Different applications have different requirements on packet loss ratio. Video is particularly sensitive to packet loss. This will be further discussed later in this chapter. Generally speaking, for real-time applications, end-to-end delay variation below 0.1 percent will ensure good service quality.

[InterQoS] recommends the following method for the measurement of packet loss ratio:

- Maximum evaluation interval = 5 minutes
- Mean packet separation = 200 ms
- The PLR metric is reported as a percentage, accurate to 0.1 percent.
- One PLR value is reported for each test period.

The Bandwidth Factor

Strictly speaking, delay, delay variation, and packet loss ratio are sufficient to determine the performance of the end users' applications, and thus the end users' QoS perception. Therefore, there is no need for bandwidth as an additional factor. However, sometimes it is easier to use the effective bandwidth available to the application to determine the performance of the application. This effective bandwidth is often referred to as throughput. For example, the bandwidth available to the transfer of a big file would determine how fast the transfer can be completed.

QOS REQUIREMENTS OF VOICE, VIDEO, AND DATA APPLICATIONS

Table 2-1 from International Telecommunication Union ([ITU]) summarizes the network performance objectives for common IP applications [Y.1541]. These objectives reflect the experts' view on the QoS need of common applications.

Note that all performance requirements listed here are one-way requirements. From the table we can see that:

1. Network delay of 100 ms would meet the need of even the highly interactive applications. Network delay up to 400 ms is acceptable for interactive applications. Roughly speaking, network delay of 100 ms is equivalent to end-to-end delay of 150 ms, which is recommended by ITU [G.114], and is commonly quoted. To a large extent, this is determined by how fast humans can react to external stimulus. Therefore, this requirement is largely applicable to all interactive applications involving humans.

2. Network delay variation below 50 ms would meet the need of even the applications with stringent delay variation requirements. Many applications don't even have an explicit requirement on delay variation.

3. Packet loss ratio below 0.1 percent would meet the need of most applications.

The following table is about network delay. Strictly speaking, it is the end-to-end delay and delay variation that matter for applications. But because end-to-end

Table 2-1 Y.1541 IP network QoS class definitions and network performance objectives

Network performance parameter	Class 0	Class 1	Class 2	Class 3	Class 4	Class 5
Delay	100 ms	400 ms	100 ms	400 ms	1 s	U^2
Delay Variation	50 ms	50 ms	U	U	U	U
Packet Loss Ratio	1×10^{-3}	1×10^{-3}	1×10^{-3}	1×10^{-3}	1×10^{-3}	U
Packet Error Ratio			1×10^{-4}			U
Targeted Applications	Real-time, highly interactive, delay variation sensitive (e.g., VoIP, Videl TeleConference, VTC)	Real-time, interactive, delay variation sensitive (e.g., VoIP, VTC).	Transaction data, highly interactive (e.g., Signaling)	Transaction data, interactive	Low loss only (short transactions, bulk data, video streaming)	Traditional applications of default IP networks

^2U means "unspecified" in this table.

delay and delay variation depend on the implementation of the end-point applications, it is more or less beyond the control of the NSPs. For example, different VoIP applications can have an end-point delay difference of 30 ms (50 ms vs. 80 ms). Therefore, presenting the delay and delay variation objectives for networks has merit. Network packet loss ratio and end-to-end packet loss ratio are the same, assuming that there is no packet loss at the end points.

The Y.1541 objectives are generalized for various types of applications. For example, Class 0 can be used for interactive voice and video. They are only concerned with the network part. In the following sections, the specific QoS requirements of voice, video, and data are discussed to give the audience a better feel. These requirements are presented from an end-to-end perspective. Note that for each application, the quality expectations can be different in different scenarios (for example, wired voice vs. wireless voice). Consequently, the requirements can also be different. We strive to provide recommendations from multiple parties to give the audience a more complete view.

End-to-End Requirements of Voice

QoS requirements for voice are well studied and understood. Most of the recommendations are fairly consistent. Based on real-world testing, Cisco recommends end-to-end performance objectives for voice-over IP [Szigeti]:

- Delay: 150 ms for high quality, 200 ms for acceptable quality
- Delay variation: 30 ms
- Packet loss ratio: 1 percent, assuming Packet Loss Concealment (PLC) is enabled
- Bandwidth: 17–106 Kbps for VoIP data plus signaling, depending on sampling rate codec, and link layer header overhead.

[Szigeti]'s recommendations are quoted here because they are based on testing in a real-world environment. They are for wired voice. For mobile voice, IEEE 802.20 Working Group on Mobile Broadband Wireless Access recommends:

- One-way network delay: <150 ms
- Packet loss ratio: <2 percent

Note that the 150 ms delay quoted in the IEEE 802.20 recommendation is network delay. Adding the end point delay, this is comparable to the 200 ms end-to-end delay in the [Szigeti] recommendation for acceptable quality. The acceptable packet loss ratio is also higher at 2 percent. In general, people accept lower mobile voice quality in exchange for mobility.

The classes applicable to voice in the Y.1541 recommendation are Class 0 and Class 1. The 150 ms end-to-end delay requirement from [Szigeti] is also consistent with Y.1541. The 30 ms delay variation from [Szigeti] is consistent with Y.1541, because it is end-to-end delay variation, and a jitter buffer at the end-point application can reduce the 50 ms network delay variation from Y.1541 down to 30 ms. The Y.1541 recommendation on packet loss ratio at 0.1 percent is more stringent because it is are intended for video applications as well.

End-to-End Requirements of Interactive Video

An example of interactive video is video teleconferencing. The QoS requirements of interactive video can vary significantly depending on quality requirement of the picture, degree of interactivity, amount of motion, and size of the picture. [Szigeti] did not recommend any explicit delay, delay variation, and packet loss ratio. Instead, it recommends provisioning of 20 percent more bandwidth over the average rate to account for burst and headers. To a certain extent, this would lead to low packet loss ratio, delay, and delay variation. It also recommends that jitter buffer be set at 200 ms.

The classes applicable to interactive video in Y.1541 are Class 0 and Class 1. Therefore, the network delay objective is 100 ms or 400 ms, depending on how interactive it is. The delay variation objective is 50 ms. Packet loss ratio objective is 0.1 percent.

End-to-End Requirements of Non-Interactive Video

Examples of non-interactive video are streaming video. [Szigeti] recommends:

- Delay: <4–5 s
- Delay variation: no significant requirement
- Packet loss ratio: <5 percent

The reason the QoS requirements of non-interactive video are very loose is large buffering can be used. Since it's not interactive, the end point can buffer several seconds of video before playing it. Consequently, the network part of delay and delay variation become almost irrelevant. Because of buffering, retransmission of lost packets is possible. Therefore, a fairly large packet loss ratio can be accommodated.

For non-interactive video on mobile devices, IEEE 802.20 Working Group on Mobile Broadband Wireless Access recommends:

- One-way network delay: <280 ms
- Packet loss ratio: <1 percent

The reason the requirements for mobile video are more stringent than wired video is the buffer and processing power on the mobile devices are much smaller. Consequently, the dependence on the network becomes higher.

The class applicable to non-interactive video in Y.1541 is Class 4. The network delay objective is 1 s, no significant delay variation objective, and packet loss ratio objective is 0.1 percent.

From these various sources, we can see that the recommendations on delay and delay variation for non-interactive video are fairly consistent. That is, delay is in the order of seconds and there is no significant delay variation requirement. The recommendations on packet loss ratio vary, largely depending on the requirement of picture quality.

To a certain extent, broadcast TV/HDTV can be classified as non-interactive video. While such broadcasts are largely real time, they can be delayed (that is, buffered) by a few seconds or minutes, as long as the entire audience sees it at the same time. In this case, there will be no significant delay or delay variation requirements on broadcast TV/HDTV. However, TV/HDTV usually has very stringent requirements on packet loss ratio, as low as 0.001 percent [Y.1541]. Y.1541 actually has two tentative classes, 6 and 7, for TV/HDTV. They are identical to Classes 0 and 1, respectively, except that packet loss ratio is 0.001 percent instead of 0.1 percent. However, it is also believed that with powerful Forward Error Correction (FEC), for example, the Enhanced DVB FEC (Digital Fountain's Raptor Code, http://www.digitalfountain.com/), 0.1 percent may work, too. Generally speaking, the exact packet loss ratio requirement of TV/HDTV requires further study.

End-to-End Requirements of Interactive and Non-Interactive Data

Examples of interactive data are web browsing, Internet chatting, etc. Examples of non-interactive data are emails, database backup, etc. Although data applications are traditionally served by the Best Effort service, their performance also depends on delay, delay variation, and packet loss ratio.

Generally speaking, interactive data applications would have good performance with delay below 150 ms. End-to-end delay of 400 ms may still be acceptable. There is generally no significant requirement on delay variation. Requirement on packet loss ratio varies, depending on the nature of the applications and whether TCP or UDP is used. If TCP is used, the applications may tolerate fairly high packet loss ratio. For example, at 1 percent packet loss ratio, many people browsing the web may not notice. Conventional wisdom among TCP researchers holds that a loss rate of 5 percent has a significant adverse effect on TCP performance, because it will greatly limit the size of the congestion window and hence the transfer rate, while 3 percent is often substantially less serious [ICFA07].

The classes applicable to interactive data in Y.1541 are Class 2 and Class 3. The class applicable to non-interactive data in Y.1541 is Class 5, which has no specific network performance objective.

PERFORMANCE OF IP NETWORKS

In the previous section, we described the QoS requirements of common voice, video, and data applications on the network. In this section, we will describe the performance of IP networks without traffic prioritization. This way, we can compare the two and find out whether there is a gap, and if so, how big. This will establish a foundation for subsequent QoS discussions, especially regarding how to provide QoS.

An IP network is said to be in normal condition when the network devices function properly and there is no congestion in the network. Essentially all IP networks are provisioned to operate without congestion, except maybe in some bottlenecks, such as peering or last-mile access links.[3] Here we also want to make it clear that there is no service differentiation in the following discussion.

An Optimistic View

With today's network devices, even at line rate or close to line rate forwarding, packets will incur a very small processing delay at each device. Virtually every

---

[3] Upgrading peering links involves contract negotiation. Therefore it can be more time-consuming. Congestion can occur during the negotiation period.

major network equipment vendor can show a third-party benchmarking result verifying that

- Delay: <1 ms
- Delay variation: <0.3 ms
- Packet loss ratio: 0 percent (or well below 10^{-6})

while forwarding at line rate.

With such network devices, a national network of air route distance up to 4190 km (U.S. diagonal, between Boston in the northeast of the United States and Los Angeles in the southwest, longer than the distance between Lisbon and Moscow) and 14 hops will have:

- Delay $= 14 \times 1 + 1.25 \times 4190 \times 5/1000 = 40$ ms

- Delay variation $\leq 14 \times 0.3 = 4.2$ ms

- Packet loss ratio: 0 percent (if a single device's packet loss ratio is 10^{-6}, then the path of 14 devices will have a packet loss ratio of about 14×10^{-6}, or 0.0014 percent. See [Y.1541] for a precise calculation method. Because it is sufficiently low, we treat it as 0 percent.

The factor of 1.25 in the above formula is used to convert the air route distance to fiber route distance, as recommended by Y.1541. The reason the device count is set at 14 will be explained in the next section. 5/1000 comes from the propagation delay of 5 ms per 1000 km. Note that all the values are one-way values, not round-trip ones. Note also that delay variation is not additive. That is, if each device has a delay variation of 0.3 ms, the whole path of 14 devices can have a delay variation between 0 ms and 4.2 ms. Here we simply use the worst-case value.

This means that a national IP network covering an area as big as the United States can easily meet the Y.1541 Class 0 requirements.

Similarly, an intercontinental network of air route distance 22,240 km (between Sydney, Australia, and Frankfurt, Germany, via Los Angeles[4]) and 24 hops will have:

- Delay $= 24 \times 1 + 1.25 \times 22,240 \times 5/1000 = 163$ ms
- Delay variation $\leq 24 \times 0.3 = 7.2$ ms
- Packet loss ratio: 0 percent

This means that an intercontinental IP network can easily meet the Y.1541 Class 1 requirements.

---

[4] People flying from Sydney to Frankfurt will likely go through Hong Kong or Singapore instead of Los Angeles, because that reduces the distance. However, packets from Sydney to Frankfurt will likely go through Los Angeles because the United States is the hub of world's undersea fiber cables, while there is little fiber connectivity across central Asia.

Most backbone NSPs would agree with the network performance statistics above. In fact, Global Crossing, a major global NSP, has a web site where people can find out network delay between any two major cities in the world:

http://www.globalcrossing.com/network/network_looking_glass.aspx

According to this web site, the reported round-trip network delay between Boston and Los Angeles is 72 ms, or 36 ms one way. The reported round-trip network delay between Sydney and Frankfurt is 315 ms, or 157 ms one way. Packet loss ratio is 0 percent. These statistics are fairly close to our simple calculation results above.

However, the following points must be noted for the network performance discussion above:

1. The per device delay, delay variation, and packet loss ratio values are for high-performance network devices. While this is applicable to pure backbone providers like Global Crossing, it is not applicable to NSPs with broadband access infrastructure because access devices are usually slower.

2. The transmission delay for access links is not included. Therefore, this discussion is not applicable for networks with low-access speed, for example, broadband access networks.

3. The possible congestion at the peering or transit is not considered. Such congestion may cause long delay, delay variation, and packet loss.

4. The delay and delay variation of the end-point applications are not considered.

Therefore, the performance statistics presented in this section represent an optimistic view of network performance.

In the next section, we present a conservative view of end-to-end performance. Because the real end-to-end performance will be somewhere in between, this approach will give us a good idea of what the real performance may be like.

A Conservative View

In this section, we first discuss delay and delay variation, then packet loss ratio.

End-to-end delay and delay variation

With some assumptions described below, ITU-T Y.1541 gives some estimate of delay and delay variation for various network devices (Table 2-2).

The link speed of an access router is assumed to be around 1.5 Mbps, for a T1 line or DSL line. Largest packet size is assumed to be 1500 bytes while VoIP packet size is assumed to be 200 bytes. At this speed, the transmission delay of a 1500-byte packet is around 8 ms. In other words, the bulk of the access router's 10 ms delay comes from the transmission delay. If the access link speed is raised to 5 to 10 Mbps, its delay and delay variation could be reduced to half or less. Similarly, the link speed of the distribution, core, and internetworking routers is assumed to be around 155 Mbps (OC-3 or STM-1). If the link speed is raised to 1 Gbps or 10 Gbps, the delay and delay variation could be reduced to a 0.5 ms level or less. Therefore, increasing link speed could be the most effective way to reduce delay and delay variation.

Table 2-2 Y.1541 – Examples of typical delay contribution by different routers

Role delay variation	Average total delay	Delay variation
Access device	10 ms	16 ms
Distribution (a.k.a. aggregation) device	3 ms	3 ms
Core device	2 ms	3 ms
Internetworking (e.g., peering) device	3 ms	3 ms

In addition to network delay, the end-point application will also add to the end-to-end delay. In Y.1541, a typical VoIP implementation is depicted below (Table 2-3).

Table 2-3 Y.1541 – Example hypothetical reference end point

Talker		Listener
G.711 encoder		G.711 decoder, with packet loss concealment (PLC)
RTP 20 ms payload size		60 ms jitter buffer
UDP		UDP
IP		IP
	Lower layer	

The encoding, jitter buffering, and packet loss concealment (PLC) described above will introduce the following end-point delay (Table 2-4).

Table 2-4 Y.1541 – End-point delay analysis

	Delay, ms	Note
Packet Formation	40	2 times frame size plus 0 look-ahead
Jitter Buffer, Average	30	center of 60 ms buffer
Packet Loss Concealment	10	one PLC "frame"
Total, ms	**80**	

To sum up all the delay factors:

$$Delay = (N_A \times D_A) + (N_D \times D_D) + (N_C \times D_C) + (N_I \times D_I) + $$
$$(1.25 \times R_{km} \times 5/1000) + 80$$

where:

N_A, N_D, N_C, and N_I represent the number of IP access, distribution, core, and internetworking devices, respectively.

D_A, D_D, D_C, and D_I represent the delay of IP access, distribution, core, and internetworking devices as presented in Table 2-4.

R_{km} is the air route distance.

80 is the end-point delay.

Therefore, this delay is the end-to-end delay.

$$\textit{End-to-end delay variation} = (N_A \times J_A) + (N_D \times J_D) + (N_C \times J_C) + (N_I \times J_I) - 60$$
$$\textit{Network delay variation} \quad = (N_A \times J_A) + (N_D \times J_D) + (N_C \times J_C) + (N_I \times J_I)$$

where:

J_A, J_D, J_C, and J_I represent the delay variation (that is, jitter) of IP access, distribution, core, and internetworking devices as presented in Table 2-4.

60 accounts for jitter buffering.

Because the delay and delay variation estimates for each network device and for the end points are conservative (that is, large), the overall estimates on end-to-end delay and delay variation are therefore conservative.

Below we present some example end-to-end delay values for some large national and intercontinental networks.

Example 1: End-to-end delay and delay variation for a U.S. national network

The assumptions made in this calculation are:

1. Distance used is approximately the span between Boston and Los Angeles, that is, 4190 km
2. Two NSPs are involved, with a total of:
   a. 2 access devices
   b. 4 distribution (i.e. aggregation) devices
   c. 6 core devices
   d. 2 internetworking (that is, peering) devices
3. Access link speed is 1.5 Mbps; other links' speed is larger than 155 Mbps.
4. Largest packet size is 1500 bytes, and VoIP packet size is 200 bytes.

$$
\begin{aligned}
\text{Delay} &= (N_A \times D_A) + (N_D \times D_D) + (N_C \times D_C) + (N_I \times D_I) + (1.25 \times R_{km} \times 5/1000) + 80 \\
&= 2 \times 10 + 4 \times 3 + 6 \times 2 + 2 \times 3 + 4190 \times 1.25 \times 5/1000 + 80 \\
&= 156 \text{ ms}
\end{aligned}
$$

This is close to the 150 ms needed for voice and other interactive applications. If the two places are not so far away, the propagation delay will be smaller so that the total delay can fall below 150 ms, too. With different end-point implementation, it is also possible to reduce the end-point delay down to 50 ms [Y.1541]. The total one-way mouth-to-ear delay would then become 126 ms.

The distribution of the end-to-end delay of the 156 ms is showed below in Table 2-5:

Table 2-5 Distribution of the end-to-end delay

	End-point delay	Propagation delay	Link delay	
	52%	16%	32%	
Percentage of total delay			Total access links 12.8%	Total non-access links 19.2%
			Per access link 6.4%	Per non-access link 1.6%

This tells us that in order to improve the delay aspect of QoS, reducing end-point delay can be the most effective approach. To reduce the delay introduced by the network path (other than propagation delay), increasing speed of the bottleneck links, which are usually the access links, will be the most effective approach. To reduce propagation delay, the caching or content delivery network (CDN) approach will be effective.

One may argue that having service differentiation at the bottleneck links may be an alternative to increasing its speed. But whether this approach will produce the same result as increasing link speed actually depends on a few factors. This will be further discussed in Chapter 8.

$$\text{Network delay variation} \quad \leq 2 \times 16 + 4 \times 3 + 6 \times 3 + 2 \times 3$$
$$= 68 \text{ ms}$$
$$\text{End-to-end delay variation} \quad \leq 2 \times 16 + 4 \times 3 + 6 \times 3 + 2 \times 3 - 60$$
$$= 8 \text{ ms}$$

This is close to the 50 ms objective of Y.1541 Class 0 and Class 1. If we take into consideration the fact that delay variation is not additive, the actual network delay variation will likely meet the 50 ms objective. The end-to-end delay variation is only 8 ms after the jitter buffer is taken into account.

Of the 68 ms of maximum network delay variation, the two access links account for 47 percent. This again points out that to reduce delay variation, increasing speed of the bottleneck links, which are usually the access links, is the most effective approach.

Another point to note in the maximum delay and delay variation calculations above is two low-speed access links are assumed, one at each end. This is applicable for consumer-to-consumer communications and possibly consumer-to-small-business communications. If one of the communicating parties has a higher-speed access link, then one low-speed access link will be replaced by one high-speed access link, and the end-to-end delay and delay variation will be reduced by about 10 ms each. On the other hand, if any access link's speed is below 1.5 Mbps, the overall delay and delay variation will increase.

Example 2: End-to-end delay and delay variation for intercontinental communications

The assumptions made in this calculation are:

1. The two communication end points are in Sydney, Australia, and Frankfurt, Germany. The air route distance between them (via Los Angeles) is 22,240 km.
2. Four NSPs are involved, with a total of:
   a. 2 access devices
   b. 4 distribution devices
   c. 12 core devices
   d. 6 internetworking (that is, border) devices
3. Access link speed is 1.5 Mbps (T1 or ADSL capacity); others' link speed is 155 Mbps or higher.
4. Largest packet size is 1500 bytes, and VoIP packet size is 200 bytes.

$$\text{Delay} = (N_A \times D_A) + (N_D \times D_D) + (N_C \times D_C) + (N_I \times D_I) + (R_{km} \times 1.25 \times 5/1000) + 80$$

$$= 2 \times 10 + 4 \times 3 + 12 \times 2 + 6 \times 3 + 22{,}240 \times 1.25 \times 5/1000 + 80$$

$$= 293 \text{ ms}$$

$$\text{Network delay variation} \qquad \leq 2 \times 16 + 4 \times 3 + 12 \times 3 + 6 \times 3$$

$$= 98 \text{ ms}$$

$$\text{End-to-end delay variation} \qquad \leq 2 \times 16 + 4 \times 3 + 12 \times 3 + 6 \times 3 - 60$$

$$= 38 \text{ ms}$$

Delay meets the [Y.1541] Class 1 requirement, but network delay variation doesn't. But again note that the delay variation value is the worst-case value. With so many (24) hops in between, the chance that all the delay variations would add up is extremely low. The end-to-end delay variation is only 38 ms.

The distribution of the end to end delay is shown in Table 2-6.

Table 2-6 Distribution of the end-to-end delay (Example 2)

	End-point-delay	Propagation delay	Link delay	
	27%	48%	25%	
Percentage of total delay			Total access links	Total non-access links
			6.8%	18.4%
			Per access link	Per non-access link
			3.4%	0.8%

Therefore, the conclusion on how to improve QoS is the same: Increasing bottleneck link speed is the most effective approach.

Example 3: End-to-end delay and delay variation for satellite communications

A single geostationary satellite can be used within the communications path and still achieve end-to-end network performance objectives on the assumption that it replaces significant terrestrial distance, multiple IP nodes, and/or transit network sections. When a path contains a satellite hop, this portion will require a delay of 320 ms, to account for low earth station viewing angle, low-rate TDMA systems, or both. In the case of a satellite possessing on-board processing capabilities, 330 ms of IPTD is needed to account for on-board processing and packet queueing delays. It is expected that most communications paths that include a geostationary satellite will achieve IPTD below 400 ms. However, in some cases the value of 400 ms may be exceeded [Y.1541].

The delay variation introduced by a satellite hop is no different from a regular wireline hop. The delay variation is mainly determined by the link speed. Therefore, the delay variation calculation approach presented in Examples 1 and 2 are applicable.

End-to-end packet loss ratio

Conservative estimates on end-to-end packet loss ratio are difficult to obtain. This is because on a link of high utilization (for example, 85 percent), while it is possible that there is no packet loss, it is also possible that a transient micro burst causes some packet loss. Because in the next section we will provide some real-world measurement results, we believe that it is acceptable to just have a qualitative discussion here.

In the networks, the potential bottlenecks are the DSLAM uplinks between the DSLAMs and the aggregation devices, the peering links between NSPs, and the last-mile access links between the customers and the NSPs. Whether packet loss will happen or not in these links is governed by a complicated dynamics. On the one hand, with the coming of video and other multimedia applications, every user is generating more and more traffic. This makes it likely for the bottleneck links to become saturated and drop packets. In addition, most web traffic is based on TCP that is "greedy," meaning that it will keep increasing the sending rate until further increasing would cause performance degradation. This could possibly cause packet loss as well. On the other hand, these factors do not necessarily cause packet loss. The bottleneck links could simply have high utilization but 0 percent packet loss. The TCP connections may terminate (for example, because a web page has been fetched) before they reach a point to cause packet loss, or TCP can be intelligent enough to slow down before packet loss occurs. Therefore, the typical packet loss ratio is not clear.

End-point intelligence, for example, packet loss concealment and TCP retransmission, can hide certain packet loss. For example, web users may not notice performance degradation at a packet loss ratio as high as 2 percent [Nessoft]. This makes it more difficult to know the true packet loss ratio—just because the user experience of web browsing is fine does not mean that there is no packet loss. The packet loss could well be there and cause problems for premium applications such as IP TV.

A Realistic View

In this section, we present some performance statistics obtained from real-world measurement. Therefore this view is a realistic view.

Figure 2-1 from [ICFA07] gives us an idea of network delay from the United States to other regions of the world. Note that the values shown are minimum Round Trip Time (RTT) values. The minimum one-way delay values will be roughly half of that.[5]

Because these are minimum values, the 95th percentile or medium or average values can be very different. Unfortunately, those are not available. Nevertheless, these statistics give an idea of real-world delay performance of today's best-effort service.

The following Service Level Agreements (SLAs) on delay and delay variation from major NSPs in the United States provide another view on real-world delay and delay variation performance [VoIPinfo]. Note that the actual delay and delay variation are usually lower than the SLA specifications, because NSPs usually want to be conservative in their SLA specifications.

- Delay
    - Qwest 50 ms maximum delay; measured actual for October 2004: 40.86 ms
    - Verio (part of NTT) 55 ms maximum delay
    - Internap 45 ms maximum delay

---

[5] One-way delay may not be exactly half of round-trip delay because the outgoing path and the returning path may be different and have different values.

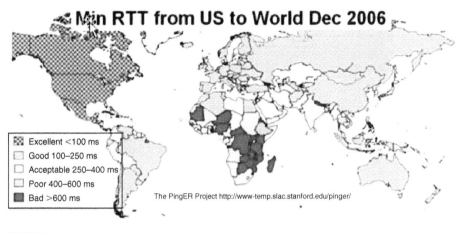

FIGURE 2-1

Minimum RTT from the United States to World in December 2006 [ICFA07][6]

- Delay variation
  - Qwest 2 ms maximum jitter; measured actual for October 2004: 0.10 ms
  - Verio 0.5 ms average, not to exceed 10 ms maximum jitter more than 0.1 percent of time
  - Internap 0.5 ms maximum jitter

It is clear from the above that delay and delay variation performance of networks in the United States meets the need of premium applications. It should be noted that the delay and delay variation values in SLAs are usually monthly average values. Therefore, there can be periods (for example, failure scenarios) when delay and delay variation are higher. Nevertheless, it implies that during normal network condition, delay and delay variation performance in works in the United States meets the need of premium applications.

Next we present some statistics on packet loss ratio.

The [PingER] project of Stanford University has done many measurements regarding packet loss in the Internet since 1998. These statistics provide much insight about the packet loss condition on the Internet, or at least from a U.S./Europe-centric perspective. Figure 2-2 shows the monthly packet loss ratio as seen from a packet loss prober at the Stanford Linear Accelerator Center (SLAC). It showed that the networks in the United States, Canada, and West Europe have a packet loss ratio close to 0.1 percent. Other regions have a higher packet loss ratio. But note that these are packet loss ratios from the prober at Stanford University (on the West Coast of the United States, a long distance away). Inside those regions, the packet loss ratio

---

[6] If the picture is not clear in black and white, the readers are encouraged to check out the original color pictures in [ICFA07].

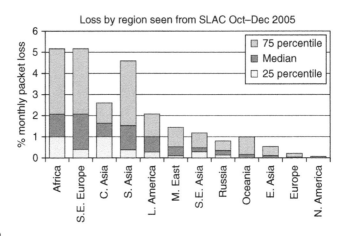

FIGURE 2-2

Packet loss ratio by region seen from SLAC [ICFAO7]

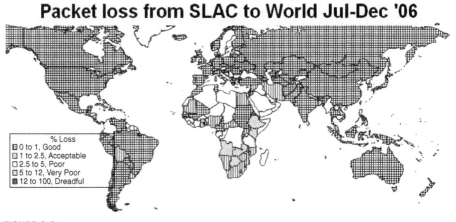

FIGURE 2-3

Packet loss condition from SLAC's perspective [ICFAO7]

should be lower. For example, inside East Asia consisting of Japan, Korea, and China, the packet loss ratio is expected to be close to 0.1 percent, too.

Figure 2-3 summarizes the network performance in terms of packet loss ratio from a SLAC perspective.

Figure 2-4 provides packet loss information about the broadband network in the San Francisco Bay Area (also known as Silicon Valley). The point to note is, since January 2003, the packet loss ratio in the broadband residential network in the San Francisco Bay Area has dropped below 0.1 percent. As most of the residents are high-tech engineers, the broadband residential network in Silicon Valley may not be a typical broadband residential network from a global perspective.

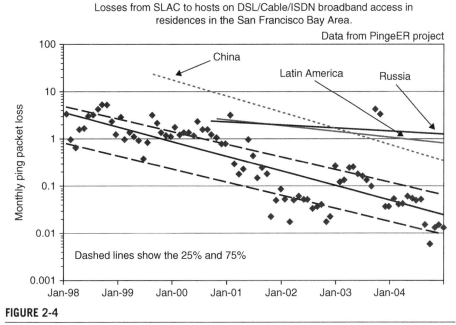

FIGURE 2-4

Packet loss ratio from SLAC to broadband access residences in Silicon Valley [ICFA07]

Nevertheless, it serves to prove that it is possible to achieve <0.1 percent packet loss ratio in broadband residential networks. The lines for China, Latin America, and Russia in the picture are for comparison purposes.

It should also be noted that this packet loss ratio was achieved before the coming of TV over broadband and the large amount of Internet video. When the amount of traffic increases dramatically, one may think that the packet loss ratio can go up. But it is not necessarily the case. This point will be discussed later in Chapter 8.

Below are some SLAs on packet loss ratio from major NSPs in the United States [VoIPinfo]. They are around 0.1 percent. Note that the actual packet loss in the network is usually far lower than the SLA specification, because NSPs usually want to be conservative in their SLA specification. For example, the measured actual loss from October 2004 is just 0.03 percent for Qwest while the SLA specification is 0.5 percent.

- Qwest 0.5 percent maximum packet loss; measured actual for October 2004: 0.03 percent
- Verio 0.1 percent maximum packet loss
- Internap 0.3 percent maximum packet loss

Again, note that packet loss ratio in SLA is usually a monthly average value. Therefore, there can be periods (for example, failure scenarios) when packet loss ratio is higher. Nevertheless, it implies that during normal network conditions, packet loss ratio in networks in the United States meets the need of premium applications.

A Seeming Contradiction in Performance Perception

Some people may be puzzled by the Internet performance statistics presented in this section. These statistics suggest that the Internet in developed countries should already be providing good service quality for applications. However, people generally feel that the Internet's performance can be unpredictable at times. How do we explain this contradiction in Internet performance perception?

This topic will be discussed in detail in Chapter 8. Here we simply drop a hint. The performance statistics presented in this section only represent the Internet's performance under normal conditions, because they are averaged. But the Internet is not the 99.999-percent-available network of which you commonly hear. It is far from that. Therefore, when citing statistics in this chapter to say that the Internet has good performance, we always quantify it as "good performance under normal conditions in developed countries."

SUMMARY

The key points of this chapter are:

- In this book, QoS means good network service quality from the end users' perspective. Anything that can affect the end users' quality perception therefore falls into the scope of QoS.

- Delay, delay variation, and packet loss ratio are the three factors that determine the end users' QoS perception.

- Different applications' QoS requirements are different. Generally speaking, end-to-end delay below 150 ms would meet the need of all known applications including the highly interactive ones. Delays up to 400 ms can be tolerable for interactive applications. Non-interactive applications have no significant delay requirement.

- Network delay variation below 50 ms would meet the need of even the delay-variation-sensitive applications. A jitter buffer can generally be used at the end point to make end-to-end delay variation meet the application's need, assuming that the end-to-end delay requirement can be met.

- Packet loss ratio below 0.1 percent would meet the need of most applications.

- Under normal network conditions, national networks as big as those covering the entire United States can meet the need of the highly interactive (that is, delay bound <150 ms) and delay-variation-sensitive applications, possibly with the aid of a jitter buffer at the end point. Intercontinental networks can meet the need of interactive applications (that is, delay bound <400 ms), except in the less-developed regions of the world where capacity is still badly lacking. The typical packet loss ratio of IP networks in developed regions can also meet the need of most applications (that is, PLR bound <0.1 percent).

- From an end-to-end perspective, the major contributors of delay are the end-point delay, possibly the propagation delay depending on distance, and the delay of the bottleneck links. The bottleneck links are also the significant contributor to delay variation and packet loss. Therefore, increasing the speed of the bottleneck link is one of the most effective approaches of providing QoS.

In this chapter, by comparing applications' QoS requirements on delay, delay variation, and packet loss ratio, and what IP networks can offer, we showed that IP networks in developed countries can meet the QoS requirements of most applications without resorting to Class of Service (CoS). This has fundamental implications on how to market QoS commercially and how to deliver QoS technically, as we will see in subsequent chapters.

Please voice your opinion about this chapter at http://groups.google.com/group/qos-challenges.

Historic Evolution of QoS Solutions

3

In this chapter, we will review the historic evolution of QoS solutions. With the QoS definition of this book, a QoS solution has two main parts: the network part and the end-point part. Again, we will focus on the network part.

To help people compare and contrast different network QoS solutions, we would like to point out that a network QoS solution is essentially a network resource allocation scheme. Each resource allocation scheme is characterized by the following aspects:

- Does it assume sufficient network resource? When there are abundant network resources, then there is no need for any explicit allocation mechanism. The answer to this question basically separates the over provisioning approach from other approaches. In reality, it is not possible that abundant resource will be available in all conditions. Therefore, this approach implicitly accepts some performance penalty when the assumption is violated, in exchange for network simplicity. In other words, this approach offers no hard performance guarantee.

- Does it do resource reservation and admission control? When resources is not abundant, there can be competition, and some resource arbitration scheme will be needed. This could be done through resource reservation and admission control, or without resource reservation. With resource reservation, resource usage can be exclusive (no sharing at any time) or resource usage can be shared if not used. Either way, there is some performance guarantee because some resource is reserved.

- How does it allocate resources among the contenders? Unless resource usage is exclusive, traffic from different sources will be mixed together in some way. To properly arbitrate, the arbiter must be able to separate different traffic groups (that is, classifying), decide how much resource each group deserves (that is, resource allocation based on SLA), and take action accordingly (that is, scheduling or policing or shaping).

37

When we discuss the various QoS solutions, we will describe how they differ in these aspects. This way, the historic evolution of QoS solutions will show us the trend in terms of these aspects.

Because only the over provisioning model assumes sufficient network resources (and simply accepts whatever quality is available when the assumption is not true), this aspect won't be explicitly discussed below.

PSTN SOLUTION

Of the communications networks, Public Switched Telephone Network (PSTN) came first. In the PSTN world, network resources are divided into pieces, each represented by a time slot. Each customer gets some fixed time slots. A connection (in other words, circuit) is set up before any communication can begin. The circuit is basically the concatenation of the time slot belonging to that customer at each network device.

PSTN circuits can be set up manually or dynamically with a protocol. When circuits are set up dynamically, to prevent the demand of resources from exceeding the supply, admission control is performed. That is, a circuit can only be dynamically set up if the admission control routine determines that there are sufficient resources in the network to meet the need. Otherwise, the circuits won't be set up. Sometimes we get a busy signal when trying to make a call. That is Call Admission Control (CAC) at work. For manually set up PSTN circuits, admission control is implicitly performed by the network operators.

To sum up, the PSTN QoS solution is characterized by two properties: (1) relying on resource reservation and admission control; and (2) allocating a resource in an exclusive way;

The pros of the PSTN solution are:

- Fixed delay.
- Zero delay variation.
- Zero packet loss (assuming the network devices function properly).

The con of the PSTN solution is:

- Resource sharing is not possible. Even if a customer is not using its time slots, those time slots cannot be used by other customers.

In summary, at the expense of potentially low resource efficiency, the PSTN solution set the benchmark for QoS solutions: fixed delay with zero delay variation and zero packet loss. As S. Keshav, computer science professor at Cornell University and a well-known QoS researcher, put it:

> *"The Holy Grail of computer networking is to design a network that has the flexibility and low cost of the Internet, yet offers the end-to-end quality of service guarantees of the telephone network." [Keshav]*

The PSTN QoS solution is widely used. Today, most of the corporate private lines are carried by PSTN circuits.

LAYER-2 QOS SOLUTIONS

PSTN networks are Layer-1 networks. Their QoS solution is a Layer-1 solution. After the PSTN networks came the Layer-2 networks such as Asynchronous Transfer Mode (ATM), Frame Relay (FR), and Ethernet. Their QoS solutions are reviewed below.

ATM QoS

ATM is also a connection-oriented technology. An ATM Virtual Circuit (VC) must be set up before communication can begin. ATM VCs can be set up dynamically with a protocol called Private Network-Network Interface (PNNI) [ATMpnni],[1] or manually by network operators. An ATM VC may or may not reserve resources, depending on what traffic it is to carry. If resource reservation is needed, then admission control will be done.

But even when resource reservation is done, there is a fundamental difference between an ATM VC and a PSTN circuit. That is, an ATM VC does not have exclusive use of network resources. Resources may be reserved for a VC. But if the VC is idle, other ATM VCs can use its resource. In fact, this is where the "virtual" word came from. From a resource allocation perspective, classification is implicitly done at the edge of the network based on the customer's interface. Some policing can also be done at the ingress of the VC to ensure that traffic will be sent at/below a certain rate. Because ATM VCs do not have exclusive use of resources, resource contention can occur. To resolve that, different VCs on a port are put into different queues. Each queue is configured with a bandwidth value. The bandwidth value is derived from the contract associated with the VC and set up either by PNNI dynamically or by network operators manually. The queues are coordinated for sending by a scheduler, which is effectively the resource arbiter. Scheduling will make sure that no VC will infringe on other VCs' resources. Therefore, no explicit policing is needed inside the network. With such a resource allocation scheme, a large number of queues and sophisticated scheduling are required. This is one of the characteristics of ATM QoS.

ATM also has other traffic management mechanisms, most notably cell discard control with Cell Loss Priority (CLP) and traffic shaping. Their usages are straightforward and won't be further discussed. Interested readers can refer to [ATMtm].

To sum up, ATM QoS solution is characterized by two properties: (1) supporting resource reservation and admission control; and (2) relying on separate queueing and sophisticated scheduling to guarantee performance while allowing resource sharing.

---

[1] In reality, PNNI is rarely used.

Armed with the above QoS solution, the ATM Forum (which was merged into the IP/MPLS Forum [IPMPLSF]) defined five ATM service categories:

- Constant Bit Rate (CBR), which supports real-time applications that request a static amount of bandwidth that is continuously available for the duration of the connection.

- Real-time Variable Bit Rate (rt-VBR), which supports real-time applications that have bursty transmission characteristics.

- Non-real-time Variable Bit Rate (nrt-VBR), which supports non-real-time applications with bursty transmission characteristics that tolerate high cell delay, but require low cell loss.

- Available Bit Rate (ABR), which supports non-real-time applications that tolerate high cell delay, and can adapt cell rates according to changing network resource availability to prevent cell loss. The ABR service category is characterized by reactive congestion control, where it uses flow control mechanisms to learn about the network conditions and adjust cell rates accordingly.

- Unspecified Bit Rate (UBR), which supports non-real-time applications that tolerate both high cell delay and cell loss on the network. There are no network service-level guarantees for the UBR service category, and therefore it is a best-effort service.

The pros of the ATM QoS solution are:

- It can provide performance guarantee and also allow resource sharing.
- There are ample service categories for users to select.

The con of the ATM solution is:

- It is very difficult to implement the ATM QoS solution. At 2.5 Gbps or higher speed, because of the need for a large number of queues and sophisticated scheduling among the queues, the hardware chips for Segmentation and Reassembly (SAR, which performs the conversion between cells and packets) and traffic management became very difficult to implement in 1998 when such high-speed interfaces became needed.

In reality, because of the popularity of IP-based applications, the most important application of ATM is to carry Best-Effort IP traffic. In other words, most of its sophisticated QoS mechanisms are rarely used.

In summary, ATM provides a very sophisticated QoS solution. It can guarantee performance and at the same allow resource sharing. But this is at the expense of high complexity, which led to high cost and lack of high-speed interface. When IP became dominant at the network layer and most IP traffic is Best Effort, ATM found itself in a unfavourable position: its cost is high while its benefit is irrelevant. This provides a good lesson on networking, regarding the relative importance of simplicity versus control (which provides some sort of guarantee but inevitably involves complexity). This is a key topic of this book and will be discussed throughout this book.

Frame Relay QoS

Roughly speaking, Frame Relay QoS can be considered a subset of ATM QoS. Frame Relay is also connection-oriented and virtual circuit based. Each Frame Relay VC has a Committed Information Rate (CIR). At the ingress of the VC, policing will be performed and nonconforming frames will be marked with a high Discard Eligibility (DE), which is very similar to ATM Cell Loss Priority. Frame Relay also has a congestion notification mechanism using the Forward Explicit Congestion Notification (FECN) and Backward Explicit Congestion Notification (BECN) bits in the frame header. Such explicit congestion notifications are supposed to notify the traffic senders to slow down when there is congestion in the network. However, because such notifications are not hooked up with the higher-layer protocol, for example, the TCP/IP stack, they are not really useful [Huston].

Frame Relay as a service rarely has its own network. Most of the time, Frame Relay is just a service interface to the users, and is carried over an ATM network. Therefore, Frame Relay does not really have its own independent QoS solution.

Ethernet QoS

In this section, in addition to discussing Ethernet QoS, we will also briefly review the development history of Ethernet itself. The reason for doing so is it will help to reveal an important point.

Ethernet was originally developed at Xerox's Palo Alto Research Center (PARC) in 1973 to 1975 by Robert Metcalfe and David Boggs. At that time, it was based on Carrier Sense Multiple Access with Collision Detection (CSMA/CD). In layman's terms, this means that before an Ethernet frame is sent, the sender will sense the carrier (that is, the cable) to see if it is idle. If it is, then the frame will be transmitted. The sender will continue to sense whether a collision happens because some other sender connected on the same cable sends at the same time. If a collision happens, then the sender will wait for a random period of time (for example, 0 to 1 s) before sending it again. If another collision happens, then the sender will wait for a longer random period (for example, 0 to 2 s) before making another attempt, and so on.

In September 1980, the 10 Mbps version of Ethernet was standardized by the Institute of Electrical and Electronic Engineers (IEEE, http://www.ieee.org). In 1993, Media Access Control (MAC) Bridging replaced CSMA/CD [802.1D]. There was no QoS notion in Ethernet until 1998 when 802.1p was published as part of the 802.1D-1998 standard. 802.1p uses a three-bit field in the Ethernet frame header to denote an eight-level priority. One possible service-to-value mapping suggested by [RFC2815] is:

Priority 0: Default, assumed to be best-effort service
Priority 1: Reserved, "less-than" best-effort service
Priority 2–3: Reserved
Priority 4: Delay Sensitive, no bound
Priority 5: Delay Sensitive, 100 ms bound

Priority 6: Delay Sensitive, 10 ms bound
Priority 7: Network Control

In reality, this mapping has not been adopted. Instead, 802.1p priority value-service mapping is locally negotiated between the setters and the interpreters, scenario by scenario.

Except by using the 802.1p field to denote a frame's priority, Ethernet does not really have any other QoS mechanism. Some were defined, for example Subnet Bandwidth Management [RFC2814], but none was seriously considered or deployed.

To sum up, Ethernet QoS is an afterthought. It is characterized by two properties: (1) not relying on resource reservation or admission control; and (2) relying on prioritization for resource arbitration.

In reality, Ethernet QoS is rarely used. To a large extent, 802.1p field, IP Differentiated Service Code Point (DSCP), and Multi-Protocol Label Switching (MPLS) Experimental (EXP) field all serve the same purpose—denoting the packet's priority. When more than one are present and one needs to be picked, MPLS EXP tends to be picked over IP DSCP over Ethernet 802.1p. Therefore, the 802.1p field is only useful in Local Area Networks (LANs) or some Ethernet-only Metro Area Networks (MANs) where packet forwarding is based on a MAC address instead of an IP address or an MPLS label. However, because LANs are lightly utilized (well below 10 percent according to [Odlyzko2]), 802.1p prioritization is rarely used there. For MANs, many NSPs are uncomfortable with Ethernet-only (that is, without MPLS switching or IP routing) solutions because of their lack of scalability. Consequently, there are few Ethernet-only MANs, and these MANs are migrating to IP or MPLS-based solution. Therefore, Ethernet QoS is rarely used in MANs either.

From the discussion above, we can see that Ethernet is simple and arguably "dumb." Theoretically, the CSMA/CD-based version cannot even guarantee that a packet can be transmitted. In fact, in March 1974, R. Z. Bachrach wrote a memo to Metcalfe and Boggs stating that "technically or conceptually there is nothing new in your proposal" and that "analysis would show that your system would be a failure." However, this simple and "dumb" technology pretty much blew away any sophisticated technologies that ever competed with it. The victims included IBM's Token Ring and Datapoint Corp. (http://www.datapointusa.com/)'s [ARCNET], which competed with Ethernet in LAN, ATM, which competed with Ethernet in MANs, and to some extent Packet over SONET/SDH (POS) in Wide Area Networks (WANs). Ethernet's winning point is exactly its simplicity, which leads to high speed and low cost albeit lack of guarantee on anything. Consistent with this point is the fact that some people took the effort to retrofit some QoS mechanisms into Ethernet, but those mechanisms are largely unused. Therefore, Ethernet contrasts sharply with ATM. One pursued sophisticated controls, got overly complicated, and failed. The other traded off guarantee for simplicity, and won. This provides another good lesson on the relative importance of simplicity versus control.

IP QOS SOLUTIONS

TCP/IP was standardized in 1981 [RFC793][RFC791]. In 1983, they replaced the Network Control Program ([NCP]) as the principal protocol of the [ARPANET]. Therefore, we can say IP networks preceded ATM networks. In the design of IP, QoS was considered. In the IP header, there was an eight-bit Type of Service (TOS) field. The definition on the TOS field is:

```
Bits 0-2:  Precedence.
Bit   3:  0 = Normal Delay,       1 = Low Delay.
Bits  4:  0 = Normal Throughput,  1 = High Throughput.
Bits  5:  0 = Normal Reliability, 1 = High Reliability.
Bit  6-7: Reserved for Future Use.


     0     1     2     3     4     5     6     7
  +-----+-----+-----+-----+-----+-----+-----+-----+
  |     |     |     |     |     |     |     |     |
  |   PRECEDENCE    |  D  |  T  |  R  |  0  |  0  |
  |     |     |     |     |     |     |     |     |
  +-----+-----+-----+-----+-----+-----+-----+-----+


Precedence
    111 - Network Control
    110 - Internetwork Control
    101 - CRITIC/ECP
    100 - Flash Override
    011 - Flash
    010 - Immediate
    001 - Priority
    000 - Routine
```

Despite the existence of the TOS field, IP QoS was not really attempted until the Integrated Services (IntServ) model was proposed in 1994 [RFC1633]. The reason was, in those days, the Internet was largely used for non-real-time applications that didn't have high QoS requirements. For example, Ray Tomlinson of [BBN] sent the first network email in 1971. By 1973, 75 percent of the ARPANET traffic was email. By 1973, the File Transfer Protocol (FTP) [RFC114] had been defined and implemented. Network Voice Protocol (NVP) was defined in 1977 in [RFC741] and then implemented, but voice over the ARPANET never worked well, largely because the links between networked nodes were not reliable. For that reason, the IP QoS mechanism was not considered helpful or attempted.

In December 1994, Netscape launched the Navigator 1.0 web browser. Its overwhelming success led people to realize that the Internet has enormous potential. The desire to use the Internet for real-time applications was rekindled.

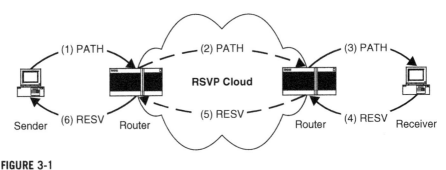

FIGURE 3-1

RSVP signaling

So was the interest for IP QoS. The effort to enable QoS in IP networks started with Integrated Services.

Integrated Services

The Integrated Services (Intserv) model [RFC1633] proposes two service classes in addition to Best Effort service. They are: (1) Guaranteed Service [RFC2212] for real-time applications requiring fixed delay bound; and (2) controlled load service [RFC2211] for applications requiring reliable and enhanced Best Effort service. The philosophy of this model is that "there is an inescapable requirement for routers to be able to reserve resources in order to provide special QoS for specific user packet streams, or flows. This in turn requires flow-specific state in the routers" [RFC1633].

RSVP was invented as a signaling protocol for applications to reserve resources [RFC2205]. The signaling process is illustrated in Figure 3-1. The sender sends a PATH message to the receiver specifying the characteristics of the traffic. Every intermediate router along the path forwards the PATH message to the next hop determined by the routing protocol. Upon receiving a PATH message, the receiver responds with a RESV message to request resources for the flow. Every intermediate router along the path can reject or accept the request of the RESV message. If the request is rejected, the router sends an error message to the receiver, and the signaling process terminates. If the request is accepted, link bandwidth and buffer space are allocated for the flow and the related flow state information will be installed in the router.

Intserv is implemented by four components: the signaling protocol (for example, RSVP), the admission control routine, the classifier, and the packet scheduler. Applications requiring Guaranteed Service or Controlled-Load Service must set up the paths and reserve resources before transmitting their data. The admission control routines will decide whether a request for resources can be granted. When a router receives a packet, the classifier will classify and put the packet in a specific queue based on the classification result. The packet scheduler will then dispatch the packets in the queues accordingly to meet their QoS requirements.

Intserv is very similar to ATM, except that it is packet-based and RSVP connections are "soft" connections. That is, RSVP connections must be regularly refreshed, or the connections will time out. Intserv represents a fundamental change to the traditional Internet architecture, which is founded on the concept that all flow-related state information should be in the end systems [Clark].

To sum up, Intserv is characterized by two properties: (1) relying on resource reservation and admission control; and (2) relying on separate queueing and sophisticated scheduling to guarantee performance while allowing resource sharing.

The pro of the Intserv QoS solution is:

- It can provide performance guarantee and also allow resource sharing.

The cons of the Intserv QoS solution are:

- The amount of state information increases proportionally with the number of flows. This places a huge storage and processing overhead on the backbone routers. Therefore, this architecture does not scale well in the Internet core.

- Ubiquitous deployment is required for Guaranteed Service. Incremental deployment of Controlled-Load Service is possible by deploying Controlled-Load Service and RSVP functionality at the bottleneck nodes of a domain and tunneling the RSVP messages over other parts of the domain.

- The requirement on routers is high. All routers must support RSVP, admission control, sophisticated classification, and packet scheduling.

In reality, Intserv was never deployed, for the reasons above. Before Intserv was fully standardized, standardization on Differentiated Services (Diffserv) had already begun. The Internet community largely prefers Diffserv over Intserv.

Differentiated Services

The essence of Differentiated Services (Diffserv) is to divide traffic into multiple classes, and treat them differently, especially when there is a shortage of resources.

IPv4 header contains a Type of Service (TOS) byte. Its meaning was described in the previous section. Diffserv renamed the TOS byte as Differentiated Services field (DS field), and defined its structure as follows [RFC2474]:

```
  0   1   2   3   4   5   6   7
+---+---+---+---+---+---+---+---+
|         DSCP          |  CU   |
+---+---+---+---+---+---+---+---+

  DSCP: differentiated services code point
  CU:   currently unused
```

DSCP denotes the forwarding treatment a packet should receive. This forwarding treatment is called the packet's Per-Hop Behavior (PHB). DSCP = 000000 denotes Best Effort treatment.

Diffserv standardized a number of PHB groups, most notably the Assured Forwarding (AF) PHB group [RFC2597] and the Expedited Forwarding (EF) PHB group [RFC2598]. Roughly speaking, the EF PHBs are intended for real-time services, whereas AF PHBs are intended for better-than-Best-Effort services (for example, Virtual Private Network, or VPN). Using different classification, policing, shaping, and scheduling rules, multiple classes of services can be provided.

Customers can mark the DS fields of their packets to indicate the desired service or have them marked by the NSP edge routers based on classification.

At the ingress of the NSP networks, packets are classified, policed, and possibly shaped. The classification, policing, and shaping rules used at the ingress routers are specified in the Service Level Agreement (SLA). The amount of buffering space required is usually not considered explicitly. When a packet enters one domain from another domain, its DS field may be re-marked, as determined by the SLA between the two domains.

Diffserv only defines DS fields and PHBs. It is the NSP's responsibility to decide what services to provide. This is both good and bad. It's good because it gives NSPs the flexibility to decide what services to provide. It's bad because this flexibility causes interoperability difficulty, as we will see in Chapters 6 and 8.

To sum up, Diffserv is characterized by two properties: (1) not relying on resource reservation or admission control; and (2) relying on prioritization for resource arbitration.

The pros of the Diffserv QoS solution are:

- Good scalability. There are only a limited number of service classes indicated by the DS field. Because resources are allocated in the granularity of class, the amount of state information is proportional to the number of classes rather than the number of flows. Diffserv is therefore more scalable than Intserv.

- Incremental deployment is possible. Diffserv-incapable routers simply ignore the DS fields of the packets and give all packets best-effort service.

- Relatively simple network operations. Sophisticated classification, marking, policing, and shaping operations are only needed at the edge of the networks. Core routers need only examine the DS field to decide the QoS treatment.

The cons of the Diffserv QoS solution are:

- There is no performance guarantee.

- Its fundamental assumption is unverified. Diffserv is only effective when there is traffic congestion and the amount of high-priority traffic is small. But how often real-world networks are like that is unverified. This topic will be revisited in Chapter 8.

■ Its effect on end users is unverified. It is envisioned that because high-priority packets receive better QoS treatment at Diffserv capable routers, their performance will be better than Best Effort. But how often can the end users perceive the difference? Little effort has been put into such verification up to this point. From a technical point of view, this may not appear important. But from a commercial perspective, this makes all the difference. This is a key topic of book and will be discussed throughout this book.

In reality, Diffserv has been deployed by some NSPs. Generally, it is not a network-wide deployment. Diffserv is only enabled in a few potential bottlenecks.

After Diffserv is introduced, there are some efforts to further extend it. The most notable efforts are MPLS support for Diffserv [RFC3270] and Diffserv-aware Traffic Engineering [RFC3564]. The former enables Diffserv to be done in a MPLS network, while the latter enables Traffic Engineering to be done per Diffserv class. MPLS support of Diffserv is deployed in some networks. Diffserv-aware Traffic Engineering has few deployments.

Hybrid IntservDiffserv

This is a hybrid model where network operators can use Intserv at the access and the edge of the network where bandwidth is more limited and scalability is less of an issue, and use Diffserv in the core of the network. [RFC2998] describes how this works, especially the treatments at the boundary of the two. However, to the best of the author's knowledge, there is lack of public awareness about this model, and there is no deployment.

Over Provisioning

The over provisioning model advocates providing sufficient capacity so that congestion will rarely occur. It is envisioned that networks like this will naturally provide good QoS. This model has a long history and can be dated back to the earliest days of the Internet.

There are some misconceptions about the over provisioning model. First of all, the word "over" has a negative implication. Because of the name, many people think that this is a naïve and somewhat wasteful way of providing QoS. They will think that such an approach is only feasible in the economic good time when NSPs have plenty of money to spend. The validity of this view will be examined later in this book. Also because of the name, many people think that this model solely relies on capacity and is against using Diffserv or other traffic management mechanisms in the networks. This is not quite true. Except for a few "fundamental over-provisionists," most people in this camp are moderate and are not against using some mechanisms for QoS. But maybe because the fundamentalists are very vocal in various organizations such as the North American Network Operators' Group (NANOG) and Internet Engineering Task Force (IETF), and these people sometimes take a brutal approach in arguing their points, they alienated a lot of people.

To sum up, the over provisioning model is characterized by two properties: (1) relying primarily on capacity for providing QoS; and (2) using other traffic control mechanisms with discretion, and not relying on resource reservation or admission control.

In reality, most IP networks have some degree of over provisioning. This will become very clear when we examine the historic utilization level of IP networks in Chapter 8. The primary reason for that is not to provide QoS, but to accommodate the rapid growth of data traffic.

TRANSPORT-LAYER AND APPLICATION-LAYER SOLUTIONS

Intelligence at the transport and application layers is very useful for providing good service quality for the applications. When QoS is narrowly scoped as using Diffserv and other traffic management schemes to provide some better-than-Best-Effort services, such intelligence may not be considered relevant to QoS. But with the QoS definition in this book, transport- and application-layer intelligence are relevant to QoS. It comprises the end-point part of a QoS solution.

Below we provide a few examples on how end-point intelligence can help the applications improve performance.

Example 1: TCP congestion control

Packet loss from network congestion can cause some received packets to become useless and create the need to retransmit even more packets. In some circumstances, this can form a vicious cycle and cause congestion collapse where the throughput of applications approaches zero while the networks become fully loaded. TCP uses a number of mechanisms to achieve high performance and avoid congestion collapse. These mechanisms include four intertwined algorithms: slow-start, congestion avoidance, fast retransmit, and fast recovery [RFC2581]. Simply put, with TCP, when a packet is received, the receiver will send an acknowledgement.[2] Therefore, if the sender gets acknowledgements back in a timely manner, the sender knows that the network and the receiver are in good condition. Therefore, the sender can increase its sending rate. In contrast, lack of acknowledgement may indicate some problems. The sender will reduce its sending rate. This simple mechanism helps TCP hunt for the optimal sending rate at the moment. It is very useful for the applications running over TCP.

---

[2] TCP is more sophisticated than that. For example, the acknowledgements of multiple packets can be combined into a single one. But for the purpose of our discussion, this simple description is sufficient.

Example 2: HTTP 1.1 persistent connection and pipelining

Today's web browsing is built upon a protocol called Hypertext Transfer Protocol (HTTP). The current version of HTTP, Version 1.1, made numerous improvements over previous versions [HTTPdif]. Here we single out persistent connection and pipelining because their effect is easier to understand.

HTTP uses TCP as its transport protocol. The original HTTP design used a new TCP connection for each request. Therefore each request incurs the cost of setting up a new TCP connection (at least one round-trip time across the network, plus several overhead packets). Since most web interactions are short (the median response message size is about 4 Kbytes [Mogul]), the TCP connections seldom get past the "slow-start" region [Jacobson] and therefore fail to maximize their use of the available bandwidth.

Web pages frequently have many embedded images. Before Version 1.1, each image was retrieved via a separate HTTP request. The use of a new TCP connection for each image serialized the display of the entire page. This made the loading of web pages with many embedded objects very slow. This contributed to Bob Metcalfe's prediction in December 1995 that the Internet would collapse in 1996.

To resolve these problems, HTTP 1.1 introduced the use of persistent connections and the pipelining. Persistent connection means that a single TCP connection can be used for multiple HTTP requests. This eliminates the need to set up many TCP connections. Pipelining means that a client can send an arbitrary number of requests over a TCP connection before receiving any of the responses. This eliminates the serialization of requests and enables faster response for web browsing.

Example 3: End-point jitter buffer

By having a jitter buffer, the end-point application can significantly reduce the end-to-end jitter (in other words, delay variation) for the application, thus providing better performance for applications that are sensitive to jitter.

Example 4: Firefox browser

For web pages with many pictures, the Firefox browser [Mozilla] loads at least twice as fast as Internet Explorer. So with exactly the same network QoS, the end-user perception on service quality is much better.

Although the focus of this book is on network QoS, it is important to point out that end-point intelligence can be very useful to enhance user experience. To some extent, this can be an alternative or supplement to network QoS, because end users will not care whether the good performance comes from network QoS or end-point intelligence. The philosophy that a network should be simple and intelligence should be at the end points is generally referred to as the "End to End Principle," originally expressed in [Saltzer]. In reality, end-point intelligence is widely used. Virtually every popular application in the Internet employs some sort of intelligence for enhancing performance.

ITU/ETSI QOS APPROACH, RACS

The International Telecommunication Union ([ITU]) and the European Telecommunications Standards Institute ([ETSI]) defined numerous standards related to QoS. In recent years, the one that is mentioned among the telecom community is the Resource and Admission Control Subsystem ([RACS]). Conceptually, RACS works as follows.

When a user needs priority treatment for a traffic stream, it would send a request for resource to the RACS server of the local domain, which maintains resource utilization information of the network. Depending on whether the resource is available, the RACS server will either grant or deny the request. Because the priority treatment may need to be done end to end, the RACS server in one domain may consult with RACS servers in other domains before granting a request. If a resource request is granted, then the RACS server will also inform the network devices along the path of the traffic stream to reserve the resource for the traffic stream. In other words, RACS is conceptually similar to Intserv [RFC1633]. However, RACS is mostly discussed in the telephony industry, in connection with IP Multimedia Subsystem ([IMS]). People from the IP world rarely talk about RACS.

FINAL OBSERVATIONS

From the discussion in this chapter, we can make the following observations:

- The sophisticated approaches for network QoS are not in use. The network industry has largely abandoned ATM and Intserv/RSVP, although they arguably offered the most advanced QoS solution that can guarantee performance and at the same time allow resource sharing. It is still too early to tell the fate of RACS. But given that networks seem to converge to IP, and IP people rarely talk about RACS, RACS's fate may turn out to be similar to ATM's and Intserv's.

- The simple approaches for network QoS are used. Of the simple approaches, the PSTN QoS approach does not allow resource sharing.[3] The Diffserv approach provides no performance guarantee. The Ethernet 802.1p approach is conceptually similar to the Diffserv approach and will be considered as a special form of the Diffserv approach. The over provisioning approach has low resource utilization and provides no performance guarantee. But because they are relatively easy to deploy and operate, they are used.

- End-point intelligence is widely used for QoS.

---

[3] In the network industry, there is currently a trend of network convergence to IP. Because of that, some people think that IP networks are simple while PSTN networks are complex. That is not the case. PSTN is indeed relatively simple compared to IP. This point will be explained in Chapter 8.

Of the surviving QoS approaches, that is, PSTN, over provisioning, and Diffserv, some appeared before the ATM/Intserv/RSVP approaches and some appeared after. Therefore, the time of appearance has no significance. Because the sophisticated approaches withered (despite that they can provide performance guarantee and allow resource efficiency), while the simple ones survive (despite lack of performance guarantee and resource efficiency), one can conclude that simplicity is more important than performance guarantee or resource efficiency. From the wide use of application intelligence, we can also conclude that it is very useful for QoS.

Of the three surviving network QoS approaches, Diffserv appears to be the mainstream approach. We say so because it is discussed most in the QoS literature. Therefore, when we discuss the commercial, regulatory, and technical challenges of QoS in this book, we will focus on the Diffserv approach.

SUMMARY

The key points of this chapter are:

- By reserving and using resources exclusively, the PSTN QoS solution can provide performance guarantee but cannot allow resource sharing.

- By using resource reservation and admission control, and separating different VCs into different queues, the ATM QoS solution can provide performance guarantee and allow resource sharing. However, this led to high complexity and its downfall.

- The Intserv and RACS approaches are similar to the ATM approach. Their fates may also be similar to ATM's fate.

- By using just packet prioritization and not using resource reservation/admission control, Diffserv can produce some QoS effect but cannot provide performance guarantee.

- By relying primarily on capacity for delivering QoS, and using traffic control mechanisms only when necessary, over provisioning appears to some people as a naïve and somewhat wasteful QoS approach, which cannot provide a performance guarantee.

- End-point intelligence is very useful for providing QoS.

- From the fate of the different QoS approaches, we can conclude that simplicity is more important than performance guarantee or resource efficiency for QoS solutions.

From the evolution of QoS solutions, Diffserv appears to become the mainstream QoS solution. Therefore, when we discuss the commercial, regulatory, and technical challenges of QoS in this book, we will focus on the Diffserv approach.

Please voice your opinion about this chapter at http://groups.google.com/group/qos-challenges.

Contemporary QoS Wisdom

4

This chapter covers the traditional thinking on how to market QoS and deliver QoS before the Net Neutrality debate started in late 2005. It should be noted that every network has its special characteristics and, therefore, different requirements on QoS. The attempt made in this chapter is to extract some common themes from many different networks so that the readers can get a high-level picture. Because of the generalization, the business model and technical solution presented in this chapter may not be directly applicable to certain networks.

BUSINESS MODEL

The network industry has been talking about QoS for more than ten years. However, to the best of the author's knowledge, no company has a well-articulated business plan to market QoS to the public. Nevertheless, from reading the marketing and technical literature, the gist of the traditional business model for QoS appears to be as follows.

NSPs will sell QoS explicitly, for example, as an additional item on top of network connectivity, and let users decide whether to buy it.

This model was exemplified by the remark of a Bell South executive in the *Washington Post*:

> *"William L. Smith, chief technology officer for Atlanta-based Bell South Corp., told reporters and analysts that an Internet service provider such as his firm should be able, for example, to charge Yahoo! Inc. for the opportunity to have its search site load faster than that of Google Inc." [WSHpost1]*

In other words, Bell South was to offer QoS separately from the connectivity service it is providing today. Customers like Yahoo! can decide whether to buy it. This view appears to be widely accepted or at least not challenged before the Net Neutrality debate.

The rationale behind this business model is: Internet companies that provide real-time or other premium services will desire good and consistent service quality to attract customers. This creates demand for QoS. Some individuals may also desire QoS in some circumstances, for example, while watching a live game of the World Cup Final online. Therefore, NSPs should make QoS available as a new service for these companies and individuals, thus creating a new revenue source for themselves. Such a model is, therefore, considered a win-win model for both the NSPs and the customers from the proponents' perspective.

Note that if a NSP offers service A and another service B = (A + QoS), and the NSP sells service B by explicitly touting the benefit of QoS, we consider it follows the QoS business model described here, because the selling of QoS is explicit and on top of network connectivity. For this reason, one commonly envisioned QoS business model in which NSPs may sell Gold and Silver services in addition to Best Effort, also follows this business model. Therefore, we can say that the traditional QoS business model is characterized by CoS-based pricing. In this book, we carefully distinguish QoS and CoS. CoS is just one possible way to realize QoS, from both a business model perspective and a technical solution perspective.

Under this QoS business model, there are some variants. A NSP can choose to adopt the soft-assurance model or the hard-assurance model. A NSP can also choose to sell QoS on demand or as a subscription service. Their differences are discussed below.

Soft Assurance vs. Hard Assurance

In the soft-assurance model:

- A NSP sells QoS as a separate entity on top of its regular connectivity service.

- Users buy QoS if they want better service quality than regular connectivity.

- The NSP provides some bandwidth or delay or delay variation assurance in the Service Level Agreement (SLA),[1] but there is no hard guarantee that such SLAs will always be met. Generally, there is an explicit or implicit agreement between the NSP and its customers that if the service quality does not meet the SLA for a period of time, service in that period of time will be free.

Therefore, the soft-assurance model is characterized by the lack of service quality assurance during abnormal network conditions and by the benign penalty when SLA is not met. For applications that need predictable service quality at all time, for example, tele-surgery or teleconferencing among VIPs, the soft-assurance model will not be that useful.

So where can the soft-assurance model be useful? It can be useful in the following two scenarios:

First, the soft-assurance model can be useful if NSPs can create better service quality with QoS that is user-perceivable. The key phrase here is "user-perceivable in normal network condition. Because soft assurance can't provide any assurance in

---

[1] To be exact, it should be in the Service Level Specification (SLS), part of the SLA. But SLA is a better-known term. In this book, we will not differentiate SLA and SLS.

abnormal network condition, it has to make a perceivable difference in normal network condition." Simply classifying traffic into different classes and giving them different PHBs may not be sufficient to create the "user-perceivable" difference, because:

- the packets may travel through other networks that do not differentiate these packets, which reduces the overall differentiation, or

- there is no congestion along the path, causing traffic prioritization to become meaningless.

Therefore, having multiple classes of services based on Diffserv may not always be an effective way to create the "user-perceivable" difference.

Second, it may be useful to sell QoS on demand. It is envisioned that an NSP providing QoS on demand will provide a web portal, whereby individual consumers can click on a "Turbo" button to buy QoS when they need it. If the soft-assurance QoS makes a difference, then the users will use it and pay for it. Otherwise, they can stop using it at any time.

From the business perspective, the advantage of this model is:

- NSPs get an additional revenue source without too much liability.

The disadvantage is:

- It is not much different from today's regular connectivity service. Today's connectivity service already provides an SLA, and there is usually some refund of service fee when the SLA is not met. Therefore, the soft assurance model may not be attractive enough for people to spend the extra money for QoS.

Now let's move on to the hard-assurance model. In this model, NSPs strive to provide predictable service quality "all the time." Compared to the soft-assurance model, QoS becomes more justifiable to pay for from the user's perspective. The problem is what kind of assurance can be given under abnormal network conditions when anything can happen? For example, a large-scale DoS attack can bring down a portion of the network, a major fiber cut can isolate part of a network, a malfunction at an edge device can cause service outage for the users connected to that edge device. It is unrealistic to expect that service quality will not suffer under such catastrophic condition. Therefore, the hard-assurance model usually takes the form of high network availability, for example, the network service will be good for the users' applications 99.999 percent of the time, combined with a large refund if the SLA is not met. The NSPs then strive to prevent the catastrophic condition from happening so that the SLA will be met.

From the business perspective, the advantages of this model are:

- By providing hard assurance of service quality and network availability, it becomes easier to convince consumers and business to buy QoS.

- High availability becomes the differentiator. NSPs need not necessarily differentiate other service characteristics such as delay and delay variation under normal network conditions.

The disadvantages are:

- Providing hard assurance can be commercially risky for NSPs because of the high penalty.

- Providing predictable service quality "all the time" is technically challenging. Such challenges will be discussed in Chapter 8.

To the best of the author's knowledge, currently no NSP provides hard assurance. This will become clear when we examine typical SLAs from the NSPs in the next chapter.

On-Demand QoS vs. Subscription-Based QoS

Another aspect of a QoS business model is whether QoS should be sold as a subscribed service with a monthly fee, or as an on-demand service. The conventional QoS business model does not prefer or preclude either one. In general, hard QoS assurance benefits all users at all times. Therefore, hard assurance is envisioned to be sold as a subscription service. Soft assurance is generally sold as an on-demand service, so that users can decide themselves at any time whether to buy or stop buying. However, if one can create a persistent and perceivable difference under normal network condition, then the soft assurance can also be sold as a subscription.

TECHNICAL SOLUTION

In the previous section, we reviewed the conventional business model of QoS. In this section, we will examine the current "mainstream technical solution." The reason we call it "mainstream technical solution" is twofold. First, this solution is most commonly seen in the QoS literature. This is where the mainstream word comes from. Second, there are different views on how QoS should be delivered, but those approaches are not explicitly expressed in the QoS literature. To reflect that, we put the term in quotations.

Traditional technical QoS solutions aim to enable NSPs to provide one or multiple higher CoS than Best Effort. For IP networks, there are basically two approaches: the Intserv/RSVP approach that is characterized by resource reservation, and the Diffserv approach that is characterized with traffic prioritization. As we discussed in Chapter 3, the Intserv/RSVP approach for the Internet was explored around 1997 but was later abandoned for lack of scalability. Diffserv has taken over as the mainstream approach to enable CoS. Therefore, a QoS solution equates to using various traffic management mechanisms to create the Per-Hop Behaviors (PHBs) defined in the Diffserv architecture. However, PHBs are mechanisms to create multiple classes of service, not the services themselves. In other words, Diffserv does not tell NSPs what CoS to provide. That's left for the NSPs to decide. This has major business implications, for example, how NSPs ensure that their CoS are compatible with each other, as we will see later.

A Common Technical Solution

If we summarize the QoS approaches published so far, a technical QoS solution will look like the following:

- At the ingress network device
    - Classify packets and mark their class at the input interface. Alternatively, a NSP can let their customers classify and mark their own packets.
    - Police the traffic. Policing can result in forwarding the packet as is, or discarding the packet, or re-marking the packet, for example, with a higher discard eligibility. Generally speaking, SLA-conforming packets are forwarded as is; SLA non-conforming real-time packets are discarded to avoid affecting performance of other real-time traffic; other non-conforming packets may be re-marked with higher discard eligibility. When re-marking happens, it is generally desirable not to change the DSCP field of the customer's packets directly. Instead, an additional header such as a MPLS header or an IP header may be added, and the EXP or DSCP field of the added header is used for the re-marking and for denoting QoS treatment inside the NSP's network. This way, when the packets get to the other side of the customer domain, their original QoS marking could be used for customer-specific QoS treatment. In other words, the customer's packets are tunneled through the NSP's network. Instead of policing, shaping could also be used at the ingress network device. Shaping causes traffic to conform to a certain rate.
    - At the egress interface, apply Random Early Detection ([RED]) or Weighted Random Early Detection ([WRED]) to decide whether to drop the packet to reduce the probability of tail drop. If not, put the packet into the appropriate queue. This is called class-based queueing. Dispatch packets from all queues are associated with an output interface according to certain rules. This is called scheduling.

- At the intermediate network devices and the egress network device:
    - Usually, there is no policing at the ingress interface.
    - At the egress interface, apply WRED/class-based queueing/scheduling as appropriate (just like at the egress interface of the ingress device).

These are illustrated in Figures 4-1 and 4-2.

Each of the steps is discussed in detail next.

Traffic can be classified in many different ways. But generally speaking, classification takes one of the following approaches:

1. Classify traffic based on the user's application's need.
2. Classify traffic based on the price that the user pays.

With the first approach, traffic in the same class has similar characteristics and requires similar handling. For example, traffic can be classified as voice, video, or data (e.g., web traffic). Diffserv can support up to eight traffic classes. Therefore, some NSPs further divide video into interactive video, which is delay and delay variation

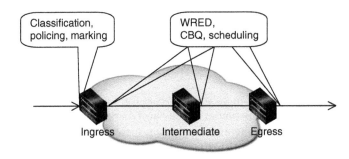

FIGURE 4-1

Traffic management in the network

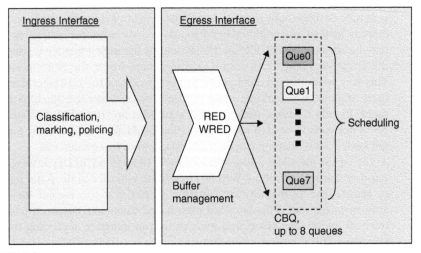

FIGURE 4-2

Traffic management mechanisms at the ingress and egress interfaces of a network node

sensitive, and non-interactive video which is not time sensitive and can be buffered. Similarly, data can be further divided into interactive data (for example, web browsing) or non-interactive data (for example, emails). Another way to classify traffic with the first approach would be to classify traffic as real-time, non-real-time business, or Best Effort. In this case, voice and interactive video can belong to the real-time class, business data services (for example, VPN) can belong to the non-real-time business class, and email can belong to the Best-Effort class. Technically, this kind of classification is based on a combination of the following fields: ingress interface; source and destination IP addresses; source and destination port numbers; type of service (TOS), which conveys the importance of each packet from the user's perspective if set; and protocol ID, which indicates which protocol each packet is associated with.

In the second approach, packets in the same class have similar monetary value to the NSPs. Their corresponding applications do not necessarily have the same characteristics or require similar service quality. The Olympic model is common, in other words, traffic is classified as Gold, Silver, and Bronze. Although the Gold class would generally correspond to the real-time class, somebody's email packets can be put into the Gold class as long as the customer is willing to pay for the high service quality. Similarly, somebody's VoIP traffic can be put into the Bronze class, mixing with mostly non-interactive data traffic, if that customer is not willing to pay. Technically, this kind of classification may be based on the ingress interface. In either case, the classification rules may come from a policy server.

What end users care about is that their applications perform satisfactorily. Therefore, they would want traffic to be classified with the first approach. This principle also makes the overall resource usage most efficient for the applications. NSPs want to make the users happy, but they also want to maximize their monetary return. Therefore they would want traffic to be classified with the second approach. When QoS is sold explicitly and users have an option of whether to buy it, it is inevitable that some VoIP users who supposedly need QoS would not be willing to pay for it and therefore will not get it, while some paying users will get QoS for their unimportant applications. In other words, the users' need and the NSP's need cannot be fully aligned. This is a commercial challenge for the NSPs. All the commercial challenges of the current QoS business model will be discussed in Chapter 6. For now, let's just assume that traffic is classified one way or another. The classification result is written to the DSCP field of the packets at the network ingress device so that other network devices can tell which class a packet belongs to and process it accordingly.

Between a user and its NSP, the SLA will specify how much traffic the user can send in each class. The SLA will also specify what the NSP should do if the user sends more than it signs up for. The SLA may dictate that the NSP accept the excess traffic and bill the user, or it may dictate that the NSP police or shape the traffic. With the former agreement, the NSP will not do policing or shaping but just do accounting. With the latter, policing or shaping can be done. Policing will cause excess packets in certain classes to be dropped or be marked with a higher discard probability, just in case discard becomes necessary later at some network nodes. An intuitive way to understand the marking of different discard probabilities would be to think that certain traffic classes have two groups per class, an "In" group for packets conforming to the SLA and an "Out" group for excess packets. Out packets will be marked with a higher discard probability. The discard eligibility marking is done on the DSCP or EXP field, etc. Intuitively those fields can be considered to have two parts: a priority class part and a discard eligibility part. The priority class part will denote which queue the packet should be put into, whereas the discard eligibility part will denote the drop probability. There can be more sophisticated policing and marking schemes, but the details are not essential to the theme of this book. In contrast, shaping will delay certain packets so that the overall traffic conforms to the rate specified in the SLA.

After traffic is classified and possibly policed, packets can be put into the appropriate queues based on the class part of their DSCP field. Each queue is characterized by three parameters:

- The (output rate/input rate) ratio
- The buffer size
- The WRED discard probability curve

While output rate of a queue can be set by network operators, the input rate may not be determined as easily. At an ingress device, from the SLAs of all customers attached, it is relatively easy to derive the upper bound of the input rate of a queue at a particular interface—all customers' traffic can be summed up. However, the instantaneous input rate can still vary greatly. The input rate of a queue at a device in the middle of a network can be more difficult to determine because after the ingress device, packets can head in different directions at different times. In addition, a network link/node failure can also change traffic distribution in the network. Therefore, input rate has to be measured regularly in the middle of the network. There, because of aggregation of many users, barring unexpected events, the input rate at a queue at a particular time of a particular day of the week is relatively stable statistically. Therefore the measured rate is useful. The ratio of output rate to input rate will determine the average queueing delay, that is, how long a packet has to wait in the queue. The higher the ratio, the lower the queueing delay. A ratio of 1.3 or higher means that when a packet arrives, the queue will be mostly empty and, therefore, the queueing delay is minimal [RFC2598]. If most delay values are small, then the delay variation values will be small, too. Such a queue is suitable for real-time traffic that is sensitive to delay and delay variation. In contrast, queues for Best-Effort traffic can be provisioned with a smaller ratio.

It should be noted that:

1. The input rate measurement must be done by the NSPs. There is little instrument or even guideline from the network equipment vendors on how to do it.

2. At an ingress device, because of lack of aggregation, the input rate can vary at any time. Measurement is therefore not useful. The derived upper bound is used instead.

3. A queue can only be configured with an output rate, not the output/input ratio. What NSPs can do is to configure the output rate based on the measured input rate and the desired ratio.

Because of the difficulty in knowing the accurate input rate, class-based queueing is generally not done except at the ingress devices. When it's done beyond that, the way the output rate of the queues is set is so arbitrary that its effectiveness is questionable. However, network congestion happens more at the edge of the network. Therefore, this approach can still be useful. This means that even the Diffserv approach will need to rely on sufficient capacity in the core to be practically useful.

Note that at the network edge, because the upper bound of the input rate is used, at any instant, the real output/input ratio can be much larger than the desired ratio, even for Best-Effort traffic. If that happens, it will reduce the differentiation between the Best-Effort class and the real-time class, because in either case, an arriving packet will be served as soon as it arrives at the queue.

Now let's move on to discuss buffer size. The buffer size determines the maximum queueing delay of a queue. A larger buffer can hold more packets before dropping them. Therefore, it allows lower packet loss ratio. However, a packet at the end of the buffer will have to wait for all the preceding packets to be transmitted before it can be transmitted. Therefore, it can have a long queueing delay. Note that the buffer size does not determine the average queueing delay. That is determined by the (output rate/input rate) ratio. The buffer size only determines the maximum queueing delay. Usually, real-time traffic cannot tolerate a large queueing delay. A VoIP packet will become useless if it is delivered too late. Therefore, the queues for real-time traffic, especially for voice, generally have a small buffer. The output/input ratio is large so the queue will be mostly empty (in other words, the buffer will be idle) anyway. In contrast, data traffic may not be that sensitive to delay, and the output/input ratio is lower. Therefore, the queues for data traffic usually have a large buffer.

For certain queues, WRED may be enabled. The goal is to drop all Out packets before any In packet is dropped. WRED ensures some fairness among different users so that users that send excessive traffic cannot affect users that observe the SLA. WRED affect packet loss ratio but not delay and delay variation for the packets that are transmitted. Whether a packet is In or Out is denoted by the discard eligibility part of the DSCP field.

Before traffic is sent back to the customer, it can be shaped to a certain rate if so desired. This is usually to avoid overwhelming the downstream customer links/nodes that are usually slower than the NSP's links/nodes.

Theoretically, policing can also be done at the egress. But this is rarely the case. Policing is to prevent excess traffic from consuming network resource. When the packets have already got to the egress of a network, this becomes moot. However, there can be special occasions that justify policing.

Conceptually, classification, policing/marking, shaping, and queue scheduling happen sequentially. That provides more flexibility because subsequent actions can make use of the result of preceding actions. But when it comes to implementation on network devices, to avoid using too many processing cycles on a packet, certain actions may be combined. This reduces flexibility. Moreover, because of different hardware architecture, different equipment vendors may implement these mechanisms differently. This can cause interoperability difficulty, a topic that will be discussed later.

In recent years, a new QoS mechanism called Hierarchical QoS (H-QoS) became popular. The background is with triple-play service, it is considered useful to be able to offer a user a certain amount of bandwidth, for example, 10 Mbps, where:

- up to 500 Kbps can be used for VoIP,
- up to 8 Mbps can be used for video, and

■ if voice and video don't use up their bandwidth, all remaining bandwidth (up to 10 Mbps) can be used for data.

This is hierarchical in the sense that the user's big pipe contains smaller pipes for voice and video inside. Without H-QoS, a QoS solution can either:

■ Classify traffic into different classes, and police or shape all the classes collectively to 10 Mbps. The problem is the individual amount of voice, video, and data cannot be controlled; or

■ Classify traffic into different classes, police or shape voice to 500 Kbps, video to 8 Mbps, and data to 1.5 Mbps. The problem is sharing is not possible and data cannot use the unused bandwidth of voice and video.

Another usage of H-QoS is to enable bandwidth control in the metro network from the Broadband Remote Access Server (BRAS), as showed in Figure 4-3 below. In the Introduction chapter we presented the architecture of the metro network. For traffic going to the end users, there are five potential congestion points, starting at the BRAS. If the BRAS has visibility of how much bandwidth each end user's voice/video/data sessions deserve, and how much bandwidth each link along the path has, then the BRAS can schedule accordingly to avoid congestion in the metro network. This idea came from the DSL Forum [DSLforum]. It is typical of the PSTN networking philosophy—strong control.

Network devices released after 2005 from major equipment vendors generally support H-QoS. When H-QoS is used, simply imagine that the scheduling mechanism at the BRAS becomes more sophisticated.

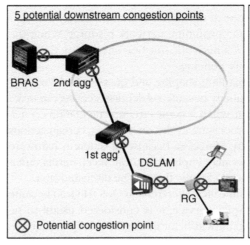
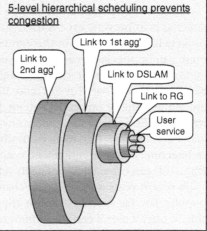

FIGURE 4-3

Hierarchical QoS illustration

In recent years, an admission control function also came into the picture. It's similar to the Call Admission Control (CAC) function in telephone networks. Sometimes callers can get a busy signal. It means that the network does not have the resource to accommodate this call. To prevent too much traffic, especially too much high-priority traffic, such as VoIP, from entering the network and causing performance degradation of all traffic, admission control is considered useful for IP networks as well. The basic concept is, if a user makes a QoS request for certain traffic on demand, the network should first figure out whether it has the resource to support the additional high-priority traffic. If it does, the QoS request will be granted and the high-priority traffic will be admitted into the network. If not, then the QoS request will be rejected so as not to affect existing high-priority traffic. The European Telecommunications Standards Institute ([ETSI])'s [TISPAN] defines Resource and Admission Control Subsystem (RACS) for this purpose. Similarly, the IP/MPLS Forum (formerly known as MPLS Forum, [IPMPLSF])'s specification on MPLS User Network Interface (UNI) includes support for such a function. A few network equipment vendors have implemented RACS or MPLS UNI. However, the author is not aware of any deployment.

Some Observations about the Technical Solution

The traditional QoS solution focuses primarily on traffic management. Other traffic control mechanisms such as routing and traffic engineering are not discussed. However, if some link or node failure causes too much high-priority traffic to concentrate on a particular link, traffic management alone cannot solve the problem. In such scenarios, an intra-domain TE mechanism may be able to offload some traffic to other links in the network, and relieve the traffic congestion. If such an intra-domain TE mechanism still cannot avoid congestion, changing the inter-domain routing policy via BGP may be able to change traffic's egress points and thus internal traffic distribution to avoid congestion. Therefore, any QoS solution that focuses on traffic management alone is inadequate. Unfortunately, existing QoS literature and solutions focus predominantly on traffic management. This is not that those authors do not understand that other traffic control schemes such as routing and TE also play an important role in determining the end user's quality perception, but that they take a traffic management-centric view on QoS. From their perspective, "QoS" means using various traffic management mechanisms to deal with traffic congestion, and "QoS" should be used with other traffic control mechanisms to deliver the Quality of Experience (QoE) that end users desire for their applications. Although there is nothing wrong technically with this perspective, the separation of QoS and QoE may not be productive because most end users don't use the QoE term. Therefore, defining QoS from an end-user perspective and equating it to QoE may be the reason why this QoS book is so different from other QoS books.

So if providing QoS to end users involves more than traffic management, a natural question is what else is involved? Just add routing and traffic engineering and the technical solution will be complete, or are there other important components?

If there are multiple components, how do they fit together? In recent years, some progress has been made towards answering these questions [Wang][Xiao1][Xiao3]. But still, the answers are not complete. For example, Content Delivery Network (CDN) and route control are not adequately discussed. Moreover, there is no discussion of the QoS business model and government regulation in existing QoS literature. This book intends to examine the commercial, regulatory, and technical challenges of QoS in depth to see if any improvement to the QoS model is needed.

SUMMARY

The key points of this chapter are:

- The traditional business model of QoS is that NSPs will sell QoS explicitly, for example, as an additional item on top of the connectivity service, and let users decide whether to buy it.

- QoS can be sold as soft assurance or hard assurance. In the soft-assurance model, there is no guarantee that the SLA will always be met. NSPs will return the service fee to the users for the time period in which SLA is not met. In the hard-assurance model, NSPs strive to meet the SLA all the time and there will be a severe penalty if the SLA is not met for a period of time.

- QoS can be sold on demand or as a subscription service. Hard assurance is envisioned to be sold as a subscription service. Soft assurance can be sold either on demand or as a subscription service.

- A common technical solution for this QoS business model is:
  - At the ingress network device's ingress interface, classify packets to the appropriate classes, then police traffic if necessary. Policing can result in discarding the SLA non-conforming packets, or re-marking them with a higher discard probability. At the egress interface, apply WRED, class-based queueing, and scheduling as appropriate.
  - At the egress interface of the intermediate network devices and the egress network device, apply WRED, class-based queueing, and scheduling as appropriate.

This traditional QoS wisdom is based on CoS. The business model is characterized by CoS-based pricing and the technical solution is characterized by traffic prioritization. In this book, QoS and CoS are two different things. CoS is just one possible way to realize QoS.

Although the traditional QoS wisdom has existed for a long time, and is accepted by many people, it has many challenges, for both the business model and the technical solution. These challenges will be discussed in subsequent chapters.

Please voice your opinion about this chapter at http://groups.google.com/group/qos-challenges.

QoS Reality

5

In Chapter 4, we discussed the traditional QoS wisdom, including a QoS business model and a technical solution. This QoS wisdom has been widely accepted by the "believers" of QoS. In this chapter, we will examine the reality of Internet QoS. This is part of the effort to check the validity of this traditional QoS wisdom. If QoS is present today, or is close to becoming reality, in the way the traditional wisdom envisioned, then the traditional QoS wisdom is likely valid. Otherwise, the wisdom may have some fundamental issues. Note again that in this book, QoS means good service quality. It may or may not be related to Diffserv or CoS.

NETWORK PERFORMANCE REALITY

While some people may have a strong opinion on this, whether the Internet already has good service quality today is a subjective matter. The perception can vary country by country, network by network, and individual by individual. Therefore, there may be no universal agreement. Nevertheless, common voice, video, and data applications' performance can be determined by delay, delay variation, and packet loss ratio. In Chapter 2, we discussed these requirements, and what the Internet can provide under normal conditions. The conclusion is, under normal network conditions, Best-Effort service in national networks as big as those covering the entire United States can meet the need of the highly interactive and delay variation-sensitive applications. Intercontinental networks can meet the need of interactive applications, except in the less developed regions. The typical packet loss ratio of IP networks in developed regions can also meet the need of most applications. All of these are achieved without introducing a higher CoS.

What does this say about the traditional QoS wisdom?

On the one hand, this says a few things unfavorable to the traditional QoS wisdom. First, if Best Effort is already good enough for most applications under normal network conditions, is there still a need for a higher CoS? This casts doubt on the traditional QoS business model whereby QoS is sold as a higher CoS. Second, this service quality is achieved without CoS or much traffic management mechanisms.

Therefore, how useful the traditional traffic management-centric solution will be for QoS is questionable, too.

On the other hand, the measured statistics on delay are minimum values and the measured statistics on packet loss ratio are monthly averages. In other words, there can be periods where delay and packet loss ratio would exceed the acceptable thresholds. In those periods, service quality can be unpredictable. For example, people who use free Internet VoIP extensively know that service quality can be very poor at times, especially for international calls. Therefore, if the traditional business model and technical solution can solve the problem in those periods, then they can still be useful. Similarly, the computations on delay, delay variation, and packet loss ratio done in Chapter 2 assume normal network conditions. Although the results appear good enough for normal network conditions, the traditional business model and technical solution can still be useful for abnormal network conditions.

Therefore, the Internet performance statistics presented in Chapter 2 cast some doubt on the validity of the traditional QoS wisdom but didn't invalidate it. These performance statistics do tell us that under normal network conditions, the traditional QoS wisdom may not be useful. Whether it will be useful for dealing with abnormal network conditions is a topic that will be discussed in Chapter 8.

The discussion above addresses the network performance aspect of QoS reality. In the remainder of this chapter, we will examine how hard real-world NSPs are pursuing the QoS business model as envisioned by the traditional QoS wisdom.

COMMERCIAL REALITY

Because the Service Level Agreements (SLAs) offered by the NSPs reveal the services that NSPs are providing, examining the SLAs is a good place to start.

SLA Is in Place

A NSP will enter a SLA when it signs up a customer, be it residential or corporate. Although some people may interpret this as meaning that NSPs are providing QoS, it is not necessarily the case. Some NSPs simply use the SLAs to limit their legal liability or impose restrictions on usage. Below we examine two real-world SLAs. One is for residential customers and the other is for corporate customers. Before we start examining the terms, we would like to caution the readers that although the terms in the SLA correlate to the degree of QoS that a NSP provides, the terms should not be interpreted as representative of the NSP's network performance. This is because NSPs usually stay conservative in the service terms to limit their liability.

SLA for residential users

SLA for residential users is usually called "Terms of Service." Below we examine the excerpt of AT&T (formerly SBC)'s residential DSL Internet Access Terms of Service. Because the SLA is rather long, we only take some parts that are relevant to our discussion here. The noteworthy clauses are italicized. The entire Terms of Service can be found at: http://edit.client.yahoo.com/cspcommon/static?page=tos. Verizon's DSL Internet Access Terms of Service is similar and can be located at: http://www.verizon.net/policies/popups/tos_popup.asp. These are the two largest NSPs in the United States.

Speed

If the Service you have purchased is provided within a range of speeds (referred to herein as the "synchronization rate" or "synch rate"), the synch rate is measured between the network interface device at your home and the equipment in the telecommunication carriers' network that provides the broadband functionality. *The Service will be provided, if at all, at least at the lowest synch rate speed within the range. There is, however, no guarantee or warranty that actual throughput from the Internet to your computer ("throughput speed") will be at or above the low end of the speed range. If the Service cannot be provided at the lowest synch rate in the range, either you or AT&T may disconnect the Service without a termination charge to you.* You will not receive a refund or rebate for the period during which you received the Service. If the Service you have purchased has an "up to" speed range (i.e. synch rate up to 384 Kbps, or up to 1.5 Mpbs), there is no minimum synch rate guarantee and there is no guarantee or warranty that the Service will perform at the upper end of the synch rate range. *Throughput speeds associated with some or all features of the Service (e.g., Usenet, email) may be limited at AT&T's sole discretion*, and such limitation will have no effect on whether the minimum speed is met.

Disclaimer of warranties

You expressly understand and agree that:

1. *Your use of the service and/or software is at your sole risk.* The service and/or software are provided on an "as is" and "as available" basis. At&t, yahoo! and their subsidiaries, affiliates, officers, employees, agents, partners and licensors expressly disclaim all warranties of any kind, whether express or implied, including but not limited to the implied warranties of merchantability, fitness for a particular purpose and non-infringement. [Note:AT&T bundles its DSL service with a Yahoo! user account. This is why Yahoo!'s name also appears here]

2. At&t, yahoo! And their subsidiaries, affiliates, officers, employees, agents, partners and licensors *make no warranty that* (i) the service and/or software will meet your requirements, (ii) the service and/or software will be uninterrupted, timely, secure, or error-free (for example but without limitation, AT&T yahoo! Does not warrant that you will always receive emails addressed to you), (iii) the results that may be obtained from the use of the service and/or software will be accurate or reliable, (iv) the quality of any products, services, information, or other material purchased or obtained by

you through the service and/or software will meet your expectations, and (v) any errors in the service and/or software will be corrected.

3. Any material downloaded or otherwise obtained through the use of the service/and or software is done at your own discretion and risk and you will be solely responsible for any damage to your computer system or loss of data that results from the download of any such material.

4. No advice or information, whether oral or written, obtained by you from AT&T yahoo! Or through or from the service and/or software will create any warranty not expressly stated in these TOS.

5. A small percentage of users may experience epileptic seizures when exposed to certain light patterns or backgrounds on a computer screen or while using the service. Certain conditions may induce previously undetected epileptic symptoms even in users who have no history of prior seizures or epilepsy. *If you, or anyone in your family, have an epileptic condition, consult your physician prior to using the service. Immediately discontinue use of the service and consult your physician if you experience any of the following symptoms while using the service—dizziness, altered vision, eye or muscle twitches, loss of awareness, disorientation, any involuntary movement, or convulsions.* [Note: This clause is not a joke. It really comes from the TOS!]

Limitation of liability
You expressly understand and agree that neither AT&T nor yahoo! Nor their subsidiaries, affiliates, officers, employees, agents, partners or licensors will be liable to you for any indirect, incidental, special, consequential or exemplary damages, including but not limited to damages for loss of profits, goodwill, use, data or other intangible losses (even if AT&T or yahoo! Has been advised of the possibility of such damages), resulting from: (a) the use or the inability to use the service and/or software; (b) the cost of procurement of substitute goods and services resulting from any goods, data, information or services purchased or obtained or messages received or transactions entered into through or from the service and/or software; (c) unauthorized access to or alteration of your transmissions or data; (d) statements or conduct of any third party on the service and/or software; (e) failure to insure the compatibility of your system (i.e., the equipment, devices, and software that you provide to receive the service) with the service and/or software, or (f) any other matter relating to the service and/or software.

Your sole remedy and exclusive remedy for any dispute with AT&T or yahoo! In connection with the service and/or software is the cancellation of your membership as provided in these TOS.

- -

The residential SLAs reveal the following things:

First, there is no mentioning of any higher CoS-enabled traffic prioritization. One possible interpretation on this is AT&T is not eager to offer such kind of QoS to residential users. Because this SLA is fairly typical among residential service SLAs, other NSPs may not be eager to do so either. This point will be revisited later.

Second, there is little assurance on service quality for residential Internet access. In particular, there is no guarantee on delay, delay variation, packet loss ratio, or service availability. But we must be careful on how to interpret this. If there is such guarantee, it means that NSPs are already providing some kind of QoS. However, lack of such guarantee doesn't mean that NSPs are not providing any QoS. What this reveals is the existence of an SLA does not necessarily mean that there is service-quality assurance, be it soft assurance or hard assurance.

Third, the SLAs serve more to limit the NSP's liability than to protect the residential users' rights.

Overall, examining the residential service SLAs serves to tell us that SLA and QoS do not necessarily have any relevance.

SLA for corporate users

Below is the corporate Internet SLA for a tier-1 NSP in the United States. The name of the NSP was replaced by a generic name "NSP1." Because the SLA is informative and not too long, and not many people have the opportunity to read the SLA between a tier-1 NSP and an enterprise, its entirety is included here. Some noteworthy clauses are italicized. This SLA is typical for corporate Internet service.

SLA.1 – Installation

Install SLA Scope. NSP1's Circuit Install SLA is to have installation of a NSP1-ordered telephone company circuit and activation of a NSP1 port completed *within forty (40) business days for T1 services, sixty (60) business days for T3 services*, and within the scheduled installation date provided in writing by a NSP1 Sales Manager for OC-3, OC-12, OC-48, FE Port Only, and Internet Dedicated Ethernet (and GigE) services.

Install SLA Process. The Install date shall be counted from the date NSP1 has received all of the following from Customer: signed contract (e.g., Service Agreement or Amendment), completed Customer Information Form, and (if requested by NSP1) completed credit application. The Circuit Install SLA is not available for Customer-ordered telephone company circuits, NSP1-ordered telephone company circuits outside the contiguous U.S., or if installation delay is attributable to Customer equipment, Customer's facility, acts or omissions of Customer, its employees or agents, Customer not passing NSP1's credit check, or reasons of Force Majeure (see below if not defined in the applicable service agreement).

Install SLA Remedy. To claim a credit, Customer must request it by calling the Billing Inquiry/Trouble telephone number on its invoice. At the time of this call, Customer must provide the company name, account number, circuit ID, Service, contact name and number, email address, SLA install date, and the actual install date in order to process the request. If NSP1 determines in its reasonable commercial judgment that there is a Circuit Install SLA non-compliance, at Customer's request, *Customer's invoice will be credited an amount equal to 50% of NSP1's billed install charge*, to include any applicable Internet Port/service install charges and NSP1-ordered and -billed access install charges for the Service for which the SLA is not compliant.

SLA.2 – Availability

Service Availability SLA Scope. NSP1's Availability Service Level Agreement (SLA) provides that the NSP1 Network (as defined in the applicable service agreement) will be *available 100% of the time.*

Service Availability SLA Process. At Customer's request, NSP1 will calculate Customer's "Network Unavailability" during a calendar month. *"Network Unavailability" consists of the number of minutes that the NSP1 Network or a NSP1-ordered telephone company circuit in the contiguous U.S. was not available to Customer, and includes unavailability associated with any maintenance at the NSP1 data center where Customer's circuit is connected or Customer's server is located other than Scheduled Maintenance (defined below).* Outages will be counted as Network Unavailability only if Customer opens a trouble ticket with NSP1 Customer support within (30) days of the outage. Network Unavailability will not include Scheduled Maintenance, or any unavailability resulting from (a) any Customer-ordered telephone company circuits or equipment, (b) Customer's applications or equipment, (c) acts or omissions of Customer or user of the Service authorized by Customer or (d) Force Majeure (see below if not defined in the applicable service agreement). If NSP1 fails to meet this SLA during any given calendar month in accordance with the above, Customer's account will be credited at Customer's request.

Service Availability SLA Remedy. To receive credit for an SLA non-compliance, Customer must request such credit within 30 days from the date of the non-compliance. *For each cumulative hour of Network Unavailability or fraction thereof in any calendar month, at Customer's request, Customer's account shall be credited for the pro-rated charges for one day of the NSP1 Monthly Fee plus one day of the telephone company line charges for the Service with respect to which a Service Availability SLA has been non-compliant.*

SLA.3 – Latency

Latency SLA Scope. NSP1's U.S. Latency SLA provides for average round-trip transmissions of *45 milliseconds or less between NSP1-designated inter-regional transit backbone routers ("Hub Routers") in the contiguous U.S.* NSP1's Transatlantic Latency SLA provides for average round-trip transmissions of *90 milliseconds or less between a NSP1 Hub Router in the New York metropolitan area and a NSP1 Hub Router in the London metropolitan area.* Latency is calculated by averaging sample measurements taken during a calendar month between Hub Routers. Network performance statistics relating to the U.S. Latency Guarantee and the Transatlantic Latency Guarantee are posted at the following location: [link omitted]

Latency SLA Remedy. If NSP1 fails to meet the Latency SLA in a calendar month, *Customer's account shall be automatically credited for that month. The* *credit will equal the pro-rated charges for one day of the NSP1 Monthly Fee* for the Service with respect to which the SLA has not been met. Credits will not be issued if failure to meet either the U.S. Latency SLA or the Transatlantic Latency SLA is attributable to reasons of Force Majeure (see below if not defined in the applicable service agreement).

SLA.4 – Network Packet Delivery

Network Packet Delivery SLA Scope. NSP1 offers both a North American and Transatlantic Network Packet Delivery SLA. *NSP1's North American Network Packet Delivery SLA provides for a monthly packet delivery of 99.5% or greater between NSP1-designated Hub Routers in North America. The Transatlantic Network Packet Delivery SLA provides for a monthly packet delivery of 99.5% or greater between a NSP1-designated Hub Router in the New York City metropolitan area and a NSP1-designated Hub Router in the London U.K. metropolitan area.*

Network Packet Delivery SLA Process. Packet delivery is calculated based on the average of regular periodic measurements taken during a calendar month between Hub Routers. Network Performance statistics relating to the Network Packet Delivery SLAs shall be posted at the following location: [link omitted]. No credits will be issued if failure to meet a Network Packet Delivery SLA is attributable to reasons of Force Majeure (see below if not defined in the applicable service agreement).

Network Packet Delivery SLA Remedy. If NSP1 fails to meet any Network Packet Delivery SLA in a calendar month, *Customer's account shall be automatically credited for that month. Such credit will equal the pro-rated charges for one day of the NSP1 Monthly Fee* for the Service with respect to which a Network Packet Delivery SLA has not been met.

SLA.5 – Denial of Service SLA

Denial of Service SLA Scope. NSP1 will respond to Denial of Service attacks reported by Customer within 15 minutes of Customer opening a complete trouble ticket with the NSP1 Customer Support. NSP1 defines a Denial of Service attack as more than 95% bandwidth utilization. This SLA is only available in the United States.

Denial of Service SLA Process. To open a trouble ticket for Denial of Service, Customer must call NSP1 at 1-800-900-0241 (Option 4) and state: "I am under a Denial of Service Attack." A complete trouble ticket consists of Customer's Name, Account Number, Caller Name, Caller Phone Number, Caller Email Address and Possible Destination IP address/Type of Attack. NSP1 shall use trouble tickets and other appropriate NSP1 records to determine, in its sole judgment, SLA compliance. Customer must notify NSP1 no later than 30 days after the Denial of Service attack(s) occurred.

Denial of Service SLA Remedy. *If NSP1 fails to meet the Denial of Service Response SLA, Customer's account will be credited, at Customer's request, the pro-rated charges for one day of the NSP1 Monthly Fee for the affected Service.*

Customer may obtain no more than one credit per month, regardless of the number of Denial of Service SLA non-compliances during the month.

SLA.6 – Reporting SLAs

NSP1 provides two types of reporting SLAs—a Network Outage Notification SLA and a Scheduled Maintenance Notification SLA. *NSP1's Network Outage SLA provides Customer notification within 15 minutes after it is determined that Service is unavailable. NSP1's standard procedure is to ping Customer's router every five minutes. If the router does not respond after two consecutive five-minute ping cycles, NSP1 will deem the Service unavailable* and the Customer's point of contact will be notified by e-mail, phone or pager, as elected by NSP1. NSP1 also will provide advance notification of Schedule Maintenance (defined below).

Scheduled Maintenance. Scheduled Maintenance shall mean any maintenance at the NSP1 hub to which Customer's circuit is connected (a) *of which Customer is notified seven calendar days in advance,* and (b) that is performed at the NSP1 hub to which Customer's circuit is connected. Notice of Scheduled Maintenance will be provided to Customer's designated point of contact by email or pager, as elected by NSP1. Upon receiving such notice, Customer may request to have such maintenance postponed to a later date if agreed to by NSP1.

Force Majeure. Any delay in or failure of performance by NSP1 will not be considered a breach of this SLA if and to the extent caused by events beyond its reasonable control, including, but not limited to, acts of God, embargoes, governmental restrictions, strikes, lockouts, work stoppages or other labor difficulties, riots, insurrection, wars, or other military action, acts of terrorism, civil disorders, rebellion, fires, floods, vandalism, or sabotage. NSP1's obligations hereunder will be suspended to the extent caused by the force majeure so long as the force majeure continues.

SLA.7 – Network Delay variation SLA (currently applicable only in U.S.)

U.S. Delay variation SLA Scope. Also known as delay variation, Delay variation is defined as the variation or difference in the end-to-end delay between received packets of an IP or packet stream. Delay variation is usually caused by imperfections in hardware or software optimization and varying traffic conditions and loading. Excessive delay variation in packet streams usually results in additional packet loss, which affects quality. *NSP1's North American Network delay variation performance will not exceed 1 milliseconds between NSP1-designated inter-regional transit backbone network routers Hub Routers in the contiguous U.S.*

Delay variation SLA Process. Delay variation shall be measured by averaging sample measurements taken during a calendar month between Hub Routers. Each month's Network performance statistics relating to the Network Delay variation SLAs shall be posted at: [link omitted]. No credits will be made if failure to meet a Network Delay variation SLA is attributable to reasons of Force Majeure (as defined in the applicable service agreement).

Network Delay variation SLA Remedy. If NSP1 fails to meet Delay variation SLA in a calendar month, Customer's account shall be *automatically credited for that month for the pro-rated charges for one day* of the NSP1 Monthly Fee for the service with respect to which Delay variation SLA has not been met.

SLA.8 – Mean Opinion Score (MOS) SLA (currently applicable only in U.S.)
Mean Opinion Score is a measure (score) of the audio fidelity, or clarity, of a voice call. It is a statistical measurement that predicts how the average user would perceive the clarity of each call

The NSP1 Internet Dedicated MOS SLA provides that NSP1's contiguous U.S. Network MOS performance will not drop below 3.8 between NSP1-designated inter-regional transit backbone network routers ("Hub Routers") in the contiguous United States. MOS is calculated using the standards based E-model (ITU-T G.107).

To receive a credit, Customer must submit their request within 30 business days after the month in which the MOS SLA was not met. Such credit will equal the pro-rated charges for one day of the NSP1 Monthly Fee for the Service with respect to the calendar month the MOS SLA has not been met.

From examining this SLA it is clear that corporate SLA is more stringent than residential ones. There are objectives on network availability, packet loss ratio, latency, and delay variation. The network availability of 100 percent, latency of 45 ms, and delay variation of 1 ms within the United States appears to be stringent, too. Plus, there is remedy for every non-compliance. However, a number of points are noteworthy.

First, there is no hard assurance of any kind. The network availability of 100 percent, packet loss ratio of 0.5 percent or lower, latency of 45 ms or lower, and delay variation of 1 ms or lower are all objectives, not hard requirements for the NSPs. If one has the opportunity to examine more corporate SLAs, one will find that this is common. Therefore, this reveals that the hard-assurance model described in Chapter 4 is unlikely to happen, at least in the foreseeable future.

Second, the remedy for every non-compliance is relatively lenient. Except for network outage, every non-compliance will cost the NSP no more than 3.3 percent (in other words, one day's worth) of monthly service fee, no matter how long the non-compliance lasts.

Third, the packet loss ratio, latency, and delay variation statistics are all monthly averages. This means that if the NSP's packet loss ratio, latency, and delay variation are so high that the user's applications are not usable for a period of 30 to 60 minutes in peak business hours, after averaging over statistics from the other 43,000 minutes in a month, the average may still be compliant. Therefore, the packet loss ratio, latency, and delay variation criteria are actually quite loose, although they appear stringent.

Fourth, all the performance criteria are for the NSP's own network only, not end to end. Because the Internet is global, the user's traffic may still suffer severe packet loss or latency or delay variation in other networks, rendering their applications not usable. However, the NSP is not accountable for that.

SLA is Loose

From the discussions in the previous section, it is safe to conclude that the terms in NSPs SLAs are fairly loose, for both residential and corporate users. The looseness has three aspects. First, it is fairly easy for NSPs to be compliant. Second, the evaluation of compliance is done by the NSPs themselves. Users have little capability to measure that. Third, in the event of non-compliance, the penalty is fairly lenient.

There Is No QoS Agreement among NSPs

Another revealing fact about QoS reality of the Internet is that there is no commercially binding QoS agreement among NSPs. A copy of a NSP peering contract is provided in Appendix A. It is concerned with Best-Effort service only. This does not necessarily mean that there is no QoS on the Internet, because Best Effort alone can still provide QoS for all applications, for example, with sufficient capacity. But this does mean that differentiating multiple CoS's will be difficult, because one NSP's different CoS's will be treated the same when they get to other NSPs' networks. Therefore, it would also be difficult to sell CoS because that requires differentiating the CoS's.

There is some effort among NSPs to set up such QoS settlement [InterQoS]. This topic will be discussed in detail in Chapter 6. Here, we simply note that it is still in an early stage, and there is no guarantee that it will be completed.

There Are Few Commercial Successes

Although virtually all NSPs claim that they provide QoS (meaning good service quality, not necessarily CoS) in their networks, to the best of our knowledge, no NSP is selling QoS explicitly for the Internet (to residential users or business connected to the Internet). By "explicitly" we mean QoS is touted to the potential users and is sold as an add-on to network connectivity, as envisioned by the traditional QoS business model.

However, there are NSPs selling QoS implicitly. By "implicitly" we mean QoS is embedded into the services that the NSPs are providing. For example, [Akamai] Technologies Inc. provides a premium Internet service by replicating certain content in its own servers, which are distributed worldwide. This effectively moves such content closer to the end users and therefore improves user experience to access such content. This is Akamai's way to provide and sell QoS for the Internet. As another example, Internap Network Services Corporation [Internap] provides a premium Internet service by connecting to multiple tier-1 NSPs, and picking the best-performing Internet path via one of these NSPs to deliver its customers' traffic. Both Akamai and Internap charge bandwidth at a higher price than other NSPs, but there is no offering of "basic bandwidth" vs. "premium bandwidth" for users to choose. QoS is effectively priced into the bandwidth and sold implicitly. There are fundamental business reasons to do so, as will be further explored in future chapters.

However, it is worth noting that little commercial success based on CoS does not necessarily translate into little Diffserv deployment on the Internet. It appears that many NSPs have some ad hoc Diffserv deployment at the edge of their networks. As to the Diffserv deployment in the Internet backbone, Bruce Davie, Cisco's Fellow specializing on QoS, stated that there is "essentially none" [Davie]. This statement was made in August 2003. The situation could be somewhat different now.

Besides the public Internet, there are also other IP networks. For example, many NSPs maintain a separate network for VPN or private lines for enterprises to interconnect their multiple sites. According to Bruce Davie, "About 200 providers running 2547 (VPN, [RFC2547])—a few dozen (of them are) doing Diffserv" [Davie]. These networks usually have some kind of CoS offering, although they may not follow the traditional QoS business model either. For example, Verizon Business's "Private IP Layer 3" service comes with six IP Classes of Service. In their own words, "Our six Classes of Service (CoS'es) let you prioritize traffic (voice, video, data) while consolidating your traffic on a single network. This offers you additional flexibility that is unmatched by most other major providers" (http://www.verizonbusiness.com/us/data/privateip/). It should be noted that what is sold is the "Private IP Layer 3" service, which comes with the six CoS's. Users can decide whether to take advantage of those CoS's, but they cannot decide whether to buy CoS's or not. Therefore Verizon's QoS offering does not exactly follow the traditional QoS business model. There are other Verizon Business IP/data services that come with CoS's, but again CoS's are not sold separately. For AT&T, its OPT-E-MAN (Optical Ethernet Metro Area Network) service offering in California provides two service options (http://www.att.com/gen/general?pid=9524&submit.y=11&submit.x=11):

- Bronze Service—supports data applications with time-varying traffic and/or those applications that are lower in priority.

- Silver Service—supports applications that require minimal loss and low latency variation, such as VoIP. Traffic in this grade of service receives a queue priority that indicates it is delay sensitive (future enhancement[1]).

This indicates that AT&T is also considering offering CoS for its corporate service network but has not made it a reality.

In summary, there is QoS selling for the Internet, but it is not based on CoS/traffic prioritization. There are some CoS offerings for corporate service networks, but CoS is still not sold separately. Therefore, we can say that the traditional QoS business model has generated little commercial success.

---

[1] As of December 2007, the web site still shows the Silver Service as "future enhancement."

SUMMARY

The key points of this chapter are:

- From a performance perspective, the Internet under normal conditions already provides fairly good service quality to common applications, in terms of delay, delay variation, and packet loss ratio. This is done without introducing a CoS higher than Best Effort. But the Internet's performance can be unpredictable at times because of failure and other abnormalities.

- From a commercial perspective, the QoS business model envisioned by the traditional QoS wisdom does not appear to be reality or close to becoming reality. Although SLAs between NSPs and their customers exist, their existence does not necessarily have relevance to QoS. SLAs are generally loose. It is fairly easy for NSPs to be compliant, users lack measurement tools for reinforcement, and the remedy for noncompliance is generally lenient. In addition, there is no QoS agreement among NSPs today. This makes differentiating among multiple CoS's very difficult because one NSP's different traffic classes will be treated the same by other NSPs. Lastly, no NSP is offering QoS commercially and explicitly today.

This chapter tells us that, if QoS means good service quality (as defined in this book), there should be no dispute that QoS is needed. However, in reality, the good enough performance of the Internet under normal conditions for applications is not enabled by the traditional Diffserv/traffic management-centric approach as envisioned by the traditional QoS wisdom. NSPs are not pursuing the traditional QoS business model in which QoS is envisioned to be sold as an add-on to network connectivity. This reality tells us that there may be some fundamental challenges with the traditional QoS wisdom. These challenges will be explored in the next part of the book.

Please voice your opinion about this chapter at http://groups.google.com/group/qos-challenges.

The Challenges

This part contains four chapters:

- Chapter 6 will discuss the commercial challenges of the conventional QoS business model.
- Chapter 7 will discuss the regulatory challenges.
- Chapter 8 will discuss the technical challenges.
- Chapter 9 will summarize the key points discussed in this part, and discuss the lessons that are learned.

The purpose of discussing the commercial, regulatory, and technical challenges is to expose the issues of the conventional QoS model. The purpose of discussing the lessons learned is to point out the direction for possible improvement of the QoS model.

To avoid possible confusion, it is worth noting again that in this book, QoS means good service quality for end users. It does not necessarily involve traffic prioritization or multiple CoS's. However, in the traditional QoS model, CoS is the incarnation of QoS, both commercially and technically. Therefore, in the context of the traditional QoS business model, QoS and CoS are the same. But generically, they are different.

Commercial Challenges

6

This chapter deals with the economic aspects of the challenges surrounding the commercialization of QoS using the conventional QoS business model. The goal is to illustrate that selling QoS as a separate item from network connectivity causes many challenges. We will discuss the challenges from the following angles:

- The challenge of deciding who should get the higher CoS.
- The challenge in selling.
- The challenge of deciding who should pay for the higher CoS.
- The challenge associated with the cooperation among NSPs for the differentiation of CoS's to end users.

Note that in the traditional business model, QoS is envisioned to be delivered with a higher CoS than Best Effort. Therefore, we refer to QoS as "a higher CoS" in this chapter.

THE "WHO SHOULD GET" CHALLENGE

The challenge here is to decide who should get the higher CoS when there is a competition for limited network resources: the customers who need it, or the customers who are willing to pay for it.

Some people may feel that the answer is clear—the higher CoS is for the customers who are willing to pay for it. If a customer is not willing to pay for it, then it doesn't matter whether his/her applications need QoS. In fact, with a fixed amount of network resources, giving QoS to the users who pay for it will likely hurt the users who need it but don't pay for it, because resource allocation is essentially a zero-sum game. However, for a NSP, putting its own interest before its customers' may not be a good way to build a successful business. It risks customer turnover. Therefore, this is a dilemma for NSPs. It would be a significant benefit if there was a new QoS business model where a NSP's interest could be aligned with its customers'.

THE "HOW TO SELL" CHALLENGE

There are multiple aspects in this challenge. They are discussed one by one below.

The "Double Selling" Difficulty

Selling is challenging. Double selling, that is, to sell first the basic network connectivity service and then QoS, is even more challenging. It is like selling an extra option or the extended warranty after selling a new car. When buying a new car, buyers are generally reluctant to buy any extra option or extended warranty that car salespeople are eager to sell. The general perception is these are not essential and you are "ripped off" if you buy them.

The "free" mentality that most Internet users have for things on the Internet will make the selling of QoS even more difficult. For historic reasons, users expect everything on the Internet to be free after they pay for the connectivity.

In developed countries, as broadband service is being used by more and more people, it is becoming a commodity. Even without the free mentality mentioned above, it will be difficult to sell an extra option on top of a commodity.

The "Evidence of Poor Quality" Difficulty

The challenge here is, in a competitive world, selling QoS on top of basic network connectivity service risks being attacked by competitors that the quality of the basic network service is poor. After all, if the basic network service is good, why would people need QoS on top of it? In other words, QoS can become the evidence of poor quality for the basic network service.

For business service, there may be an additional barrier for QoS selling. So far, Best Effort has been the only service that NSPs sell. Because of competition, while selling the Best-Effort service in the past, a salesperson might have claimed superb quality of the Best Effort service. Now the salesperson faces a dilemma. If he goes back to the customer and downplays the quality of the current service in an attempt to sell QoS, he is effectively reducing his own credibility. This risks customer turnover. If he doesn't downplay quality of the current service, the customer is unlikely to buy QoS.

Some people may think that QoS can be sold by saying that today's Best-Effort service is good for web browsing but not good enough for real-time applications. There are two problems with this approach. First, VoIP is nothing new. It has been around for many years and has flourished with just basic connectivity. The biggest video providers on the web, for example, Google (which acquired YouTube) and MSN, think that their video would do just fine with basic connectivity, too. They are against selling QoS by NSPs in the Net Neutrality debate. Second, if one further argues that today's VoIP and video are not toll quality, and toll quality VoIP and video would require QoS, then QoS would need to provide some pretty strong assurance. That's by itself a major challenge, as will be discussed in the next section.

The "What Assurance to Provide" Difficulty

As explained in Chapter 4, the soft-assurance QoS business model can be simply described as "try one's best to make a difference on service quality with QoS but make no guarantee." This is not too different from today's regular connectivity service. Therefore, it may not be attractive enough for people to pay the extra money for a higher CoS. After all, today's connectivity service already comes with an SLA. However, if one is to offer hard assurance, then there is considerable commercial risk. The challenges of both models are described below.

As we explained in Chapter 4, the soft-assurance model can be useful for the following scenarios:

- First, where NSPs can create user-perceivable differences in service quality under normal network condition.
- Second, for selling QoS on demand and letting the users decide themselves whether the soft-assured QoS makes a difference at the needed time.

The challenge for the first scenario is that under normal conditions, networks in developed countries are already capable of meeting the needs of most applications. Therefore, soft assurance can only be useful for networks in developing countries where bandwidth can be limited. However, in those countries, people and corporations may be more tolerant to marginal service quality and less willing to pay for QoS. As a result, in developing countries, CoS may become an internal cost-saving mechanism for NSPs to conserve bandwidth, not a service for selling as envisioned by the conventional business model. Consequently, in both developing and developed countries, soft assurance is unlikely to generate much revenue.[1]

The challenges for the "selling QoS on demand" scenario are, first, selling QoS on demand generally means selling QoS at a time the network is not performing at its best. Otherwise, the users would not consider buying QoS on demand in the first place. When the network is not performing well, turning some traffic into high priority or introducing additional high-priority traffic may exacerbate the problem; second, for on-demand QoS, QoS accounting and billing must also be invoked on demand, that is, in a dynamic fashion. This is more complicated than the subscription-based case in which the QoS accounting and billing is static. In summary, a considerable amount of technical effort is needed to enable QoS on demand. However, because the applications that the on-demand QoS is used for (for example, online movies or VoIP calls) don't cost much themselves, individual users will not be willing to pay much for on-demand QoS either. As a result, on-demand QoS revenue may not justify the complexity and overhead of the technical effort.

In the hard-assurance model, NSPs strives to provide good service quality at all times, coupled with a severe penalty if the SLA is not met. To a certain extent, QoS is sold as an insurance. The challenges of the hard-assurance model are outlined here.

---

[1] Note that all the discussions in this chapter are for the conventional QoS business model in which users have the option to decide whether to buy QoS.

First, it is commercially risky for the NSPs because of the high penalty for noncompliance. Incurring a penalty may wipe out revenue for a significant period of time.

Second, it is technically difficult to achieve. Today's network devices are highly sophisticated systems. Although hardware may approach 99.999 percent reliability, software reliability is generally lower, especially if the system is frequently configured and reconfigured, for example, to introduce new features. In addition, because nobody can guarantee anything under a network catastrophe, for example, when a major fiber cut happens, hard assurance is envisioned to be accomplished by doing everything to prevent a network catastrophe from happening. This means that the networks offering hard assurance will have high availability. Therefore, hard assurance and Best-Effort services are unlikely to coexist in the same network, or everyone will just use the highly available Best-Effort service. The traditional business model of selling QoS as a higher CoS is not applicable.

THE "WHO SHOULD PAY" CHALLENGE

In order for a NSP to sell QoS, it must decide whether to charge it to business or consumers or both. Because communications that would benefit from QoS may happen between two consumers, too, the NSP may also need to decide whether the calling consumer or the receiving consumer should pay. For our purposes, the calling consumer is the one that initiates the communication. Because communications are mostly bi-directional, the calling consumer can be both a packet sender and a packet receiver during the communication session. This is why we don't call this entity the "sender" but the "calling consumer" instead.

The reason we address the "who should pay" question first by "business or consumer" and then by "calling consumer or receiving consumer" is that business and consumer are perceived to have different amounts of financial resources and thus different willingness to pay for QoS. NSPs generally prefer to charge QoS to businesses instead of consumers.

Note that the paying party has the power to choose whether to pay for QoS. If the paying party chooses not to pay, QoS will not happen.

Business Pay or Consumer Pay?

Because businesses have more money and are generally less price sensitive, and consumers are exactly the opposite, NSPs generally prefer to charge QoS to businesses. In an interview with *Business Week* in October 2005, Edward Whitacre, Chairman and CEO of AT&T (then SBC) stated this preference [Bizweek].

However, that is the most opposed scenario by the Net Neutrality proponents. In his testimony to the U.S. Senate in February 2006, Lawrence Lessig, one of the major advocates of Net Neutrality, stated that:

> *"Consumer-tiering, however, should not discriminate among content or application providers. There's nothing wrong with network owners saying*

'we'll guarantee fast video service on your broadband account.' There is something wrong with network owners saying 'we'll guarantee fast video service from NBC on your broadband account.' " [NNlessig]

Here "guarantee fast video service from NBC" means charging NBC a QoS fee and giving its traffic higher priority. Net Neutrality will be discussed in detail in Chapter 7.

As of early 2007, public opinion appears to be more sympathetic to Net Neutrality than opposed to it. Therefore, charging QoS to businesses will have its challenges. Partly because of this and partly because of its desire to get quick approval for its merger with Bell South, in December 2006, AT&T committed not to charge QoS on the Internet for two years.

Ironically, it is the companies such as Google (which provides video service via its acquisition of YouTube), eBay (which provides VoIP service via its acquisition of Skype), Yahoo, and Microsoft MSN, which all supposedly need QoS most, that are leading the effort in the Net Neutrality debate to prevent NSPs from doing traffic prioritization and charging for QoS. These companies would benefit most from QoS because, first, they provide video and VoIP over the Internet, which supposedly needs QoS for good service quality; second, these companies can afford QoS; third, QoS could become a barrier against future competing startups because QoS fees have a bigger impact on the startups, which do not have as much financial resources. This fact casts doubt on the traditional QoS wisdom that if a NSP can offer QoS, there will be corporate demand for it—even the companies that can supposedly benefit most from QoS are not willing to pay for it.

While charging businesses for QoS is challenging, charging both businesses and consumers for QoS will be even more challenging. Depending on how the Net Neutrality debate develops, NSPs may be left only with the option of charging QoS to consumers. From an economic perspective, this is not very attractive.

In the unlikely event that NSPs win the Net Neutrality debate completely, the "business pay" option can still have a challenge. Businesses are usually the content providers on the Internet. A multiservice Internet may come with multicast. A business may send a single stream over its access interface into its NSP's network, which will be replicated many times in its NSP and possibly other NSPs' networks. How much should the business pay for QoS, the number of copies it sends, or the number of copies its users receive? Even if the ICPs are willing to pay a QoS fee that is based on the number of receivers, the NSP still has the following difficulties in implementing this model, because

- The NSP may need to coordinate with other NSPs to figure out the exact number of receivers in a multicast group.

- The NSP must keep a complex billing record in order to reduce potential business disputes. The reason is that multicast QoS is notoriously difficult to provide, even in theory, not to mention the constraints in vendors' specific implementations and the complexity in operations and maintenance. In short, when a traffic stream is replicated many times and each stream takes a different path, it is likely that some streams will receive better QoS than others. Some receivers may not really get the QoS that the sender intends to provide. Therefore, the NSP has to work closely with other NSPs to keep track of which receivers get the QoS and which don't, a very complicated task. Otherwise, business disputes between the ICPs and the NSPs on how much QoS is delivered may result.

Besides multicasting, Peer To Peer ([P2P]) networking poses yet another challenge for charging QoS to business. The problem is with P2P, the consumers that download the content from the business will help to distribute the content as well. However, the consumers will not be paying for QoS. With P2P, service quality for the business may become good enough. The business, therefore, may not be willing to purchase QoS.

Calling Consumers Pay or Receiving Consumers Pay?

First, let's discuss the "both calling consumers and the receiving consumers pay" scenario. This would be ideal for the NSPs. However, this is like FedEx wanting to charge both the sender and the receiver for the delivery of a package. This model will be unlikely to work because it requires that both the sender and the receiver agree to pay for QoS. This would require some signaling mechanism that does not exist today. Assuming that such a signaling mechanism exists, there is a likelihood that one or both parties are not willing to pay for QoS so the NSP will not be able to collect the QoS fee. Therefore, such a mechanism will unlikely be developed because it is a fair amount of work for an uncertain result, especially given other uncertainties such as Net Neutrality surrounding QoS. In other words, this model will be unlikely to work.

Second, let's discuss the "calling consumers pay" scenario. First, even if a consumer is willing to pay for QoS, he/she will likely choose to pay on demand. This is because consumers are generally cost-sensitive and would buy QoS only when necessary. Next, because communications between consumers are most likely bi-directional, a calling consumer can sometimes be a packet sender and sometimes a packet receiver. When the calling consumer is a packet receiver, there must be a signaling mechanism to notify the ingress network device connecting to the packet sender to prioritize packets heading to this receiver so that QoS treatment at the bottlenecks is possible. This is a complicated procedure and there is no well-defined way to signal this. If multiple NSPs are involved (which is generally the case), it becomes even more complicated. Cisco may have a proprietary mechanism based on BGP for this purpose, but to the best of the author's knowledge, this mechanism is not widely implemented by other vendors. Therefore, there will be interoperability challenges if this option is chosen.

Third, let's discuss the "receiving consumers pay" scenario. Again, because a receiving consumer can sometimes be a packet sender and sometimes a packet receiver, the above challenge applies.

Some people may think that there is a fourth option, which is to let the callers and receivers reach their own agreement regarding which party will pay for QoS, just like UPS or FedEx users can reach their own agreement regarding who will pay for a shipment. However, the value of a UPS or FedEx package is usually high enough to make such a negotiation worthwhile. On the Internet, QoS may be for an online movie or a VoIP call. The value of QoS for it generally doesn't justify a negotiation.

THE "LACK OF INTERPROVIDER SETTLEMENT" CHALLENGE

Every NSP that chooses to adopt the conventional QoS business model will face the challenges discussed in the previous sections. The decisions of how to deal with those challenges belong solely to each individual NSP. But there is also an interprovider challenge associated with the conventional QoS business model.

To make the QoS effect perceivable to the end users, it is desirable for CoS to be enabled end to end. Therefore, cooperation of all NSPs along the path is essential. This is so because if only NSP1 is differentiating traffic in its network but other NSPs are not, the potential QoS buyers would question whether that would make a sufficient difference. Consequently, they may decide not to buy QoS from NSP1. To enable the cooperation, all the involved NSPs must be able to benefit monetarily from providing QoS to a particular application session. Otherwise, other NSPs would not have incentive to give preferential treatment to the beneficiary NSP's high-priority traffic. To the best of the author's knowledge, there is no economically binding QoS settlement between NSPs as of January 2007. A typical peering contract among NSPs is included in Appendix A. There is no provisioning on QoS settlement in the contract. This is a major barrier for the commercialization of CoS.

There is some effort towards interprovider QoS. The most notable one is the QoS Working Group of the MIT Communications Futures Program [MITcfp]. This effort started in 2004. The participants include some of the brightest minds and biggest names in the QoS area. One of the key goals from the beginning is to produce an interprovider QoS white paper. In November 2006, draft 1.1 was publicly released. It is likely the most comprehensive document to date on the issues of interprovider QoS. While some possible solutions are discussed for the issues, the proposal was largely exploratory. The white paper stated at the Executive Summary that:

> *"Some providers are now beginning to interconnect with each other via 'QoS-enabled peering' in an attempt to offer QoS that spans the networks of multiple providers. However, in the absence of appropriate standards and established procedures for management, trouble-shooting, monitoring, etc., such interconnections are likely to be challenging and labor-intensive. This document seeks to identify the key issues that service providers need to agree upon if inter-provider QoS is to be readily deployable."*

The white paper further stated that:

> *"This paper has two main goals:*
>
> ■ *To identify standards that should be worked on to simplify deployment of interprovider QoS*
>
> ■ *To identify 'best common practices' that, while not requiring standardization, could ease the deployment of interprovider QoS if agreed to by a critical mass of providers"*

In other words, the QoS experts considered the industry at the "identify the key issues" stage in November 2006. "Service providers need to agree upon" these issues, and standardize the solutions and work out some best practice, before interprovider QoS can be deployable. Given that the authors of that white paper are some of the most knowledgeable and experienced QoS experts in the industry, one can conclude that interprovider QoS with the conventional QoS model (that is, QoS is sold as a higher CoS) is far from becoming reality.

A natural question that follows is, given the effort like [MITcfp], won't interprovider cooperation be just a matter a time? After all, NSPs have a common interest here. While it is true that NSPs have a common interest here, cooperating to provide QoS may have too many challenges to happen. First comes the regulation obstacle. The NSPs that have the most customers are often incumbent broadband access providers that enjoy a monopoly or duopoly status. Cooperation among them, especially with the purpose to increase price, may cause regulation scrutiny. This is something that the incumbent broadband access providers very much want to avoid. One may argue that when their survivability is threatened, NSPs will undertake this daunting task no matter how challenging it is. The question is, is QoS critical at all to NSP's survivability? Because QoS is sold as an add-on to network connectivity to enable premium services such as VoIP or video, QoS revenue is effectively a derivative revenue of voice and video revenue. Therefore, QoS revenue will likely be a fraction of VoIP and video revenue. However, VoIP is often free or marginally priced. Communications history also showed that past content distribution service such as the distribution of newspapers by the postal service, despite its tremendous volume, only generates a small part of the total revenue [Odlyzko4]. If the trend holds true (a topic that will be discussed in Chapter 10), video distribution will not generate much revenue for NSPs either. The abundance of free or low price Internet video service is an early evidence. Therefore, QoS revenue is unlikely to be big enough to become a top priority for NSPs. This may have been the reason why NSPs haven't made any serious effort towards interprovider QoS over the years.

What can be even more daunting is, even if we assume that one day NSPs cooperate to provide QoS, the amount of coordination is enormous. People can get a comprehensive picture of what coordination is needed from [InterQoS]. The following is an example of the coordination needed to set up a maintenance window.

Every NSP needs some maintenance window in which it can perform certain network maintenance, which may cause service outage. In today's Best-Effort world, scheduling a maintenance window is a fairly straightforward thing. A NSP can decide on its own which date it wants to perform the network maintenance. The maintenance window is usually in the wee hours, for example, two to four a.m., to minimize impact to its customers. The NSP will inform other NSPs ahead of time as a courtesy but will not need explicit consent from them. In contrast, to support interprovider QoS, [MITcfp] stated that:

> *"Any provider that has customers that were likely to be unreasonably impacted by another providers planned outage would have the right to negotiate changes to the requested window. In this case, any changes agreed must still be communicated in accordance with the notification period and procedures to all other affected providers. This regime would require all notification requests to be cascaded through providers as one provider may not know what it used beyond the adjacent provider's network."*

Network engineers who have to schedule maintenance windows will be shocked when they learn that the task has become so complicated. People who are not network operators can still get an idea how tedious this can be by imagining that they have to coordinate their tasks at work with people in three other companies.

In the end, the authors of the white paper summarized that:

> *"We have not yet collected and analyzed sufficient issues, practices and potential solutions to this maintenance window aspect of Interprovider QoS. Therefore we have no complete "best practice" proposal yet. This is an area for further study."*

Keep in mind that the authors of the white papers are the QoS experts in the industry, and the issue under discussion is a supposedly simple task of scheduling a maintenance window. This is just an illustration of the challenge that selling QoS explicitly can bring.

The coordination among NSPs also involves how to allocate the end-to-end delay, delay variation, and packet loss ratio budget. This is both a commercial issue and technical issue, and will be discussed in Chapter 8.

SUMMARY

The key point of this chapter is that there are many commercial challenges associated with the conventional QoS business model in which QoS is sold as a separate item from network connectivity, that is, as a higher CoS. Some of the major challenges are:

- There is a dilemma regarding who should get QoS, the customers who need it or the customers who are willing to pay for it.

- Selling QoS after selling network connectivity involves two sellings. It's more difficult than selling just once.

- Competition makes it difficult to sell QoS on top of network connectivity, because competitors can attack the quality of the basic service.

- Both soft- and hard-assurance models have challenges. Soft assurance cannot provide a strong guarantee on service quality. It therefore may not be attractive enough for people to buy QoS. Hard assurance is technically difficult and commercially risky for NSPs.

- NSPs would like to charge QoS to businesses such as Internet content providers (ICPs). However, the Net Neutrality controversy may prevent that from happening. Even without Net Neutrality, multicast and P2P technologies will introduce challenges for charging QoS to business. Charging QoS to consumers is less attractive economically and may require a signaling mechanism, which doesn't exist today for the purpose of prioritizing traffic at the network ingress.

- There is a lack of interprovider QoS agreement. This hampers NSPs from cooperating on QoS and makes it difficult for NSPs to commercialize QoS. Even if interprovider QoS agreements exist, the coordination required among NSPs would be complicated and costly.

Tackled individually, every issue discussed in this chapter is likely resolvable. But some of the solutions can be complicated. However, commercializing QoS would require that most of them be resolved. This makes the task daunting and discourages NSPs from springing into action. Combined with the regulatory and technical challenges that will be discussed in the next two chapters, the task of commercializing QoS based on the conventional business model seems daunting and unattractive.

Fundamentally, the explicit selling of QoS is the cause of the difficulties. By selling QoS explicitly, NSPs have to tout its benefits. This raises the expectation of the users and sets up the NSPs to face all kinds of challenges and scrutiny. The commercial challenges are discussed in this chapter. The regulatory and technical challenges will be discussed in the next two chapters.

Please voice your opinion about this chapter at http://groups.google.com/group/qos-challenges.

Regulatory Challenges

7

From the first day the concept of QoS was conceived, it's always taken for granted that if a NSP can provide QoS, it will have the right to sell QoS. Well, that's before the Net Neutrality debate started. In this chapter, we will discuss what Net Neutrality is and how it affects the traditional QoS business model.

THE NET NEUTRALITY DEBATE

Arising of Net Neutrality

The precise definition of network neutrality varies. Generally speaking, a network is neutral if it has no restrictions on what kinds of equipment can be attached, has no restrictions on whether or how equipment can communicate, and does not degrade one set of communications for the sake of another. From a QoS angle, Net Neutrality is a debate regarding whether NSPs should be allowed to prioritize and treat Internet traffic differently based on whether a QoS fee is paid.[1]

In an interview with *Business Week* in November 2005, when asked "How concerned are you about Internet upstarts like Google, MSN, Vonage, and others?," AT&T Chairman and CEO Edward Whitacre, Jr. responded:

> *"How do you think they're going to get to customers? Through a broadband pipe. Cable companies have them. We have them. Now what they would like to do is use my pipes free, but I ain't going to let them do that because we have spent this capital and we have to have a return on it. So there's going to have to be some mechanism for these people who use these pipes to pay for the portion they're using. Why should they be allowed to use my pipes? The Internet can't be free in that sense, because we and the cable companies have made an investment and for a Google or Yahoo! or Vonage or anybody to expect to use these pipes [for] free is nuts!" [Bizweek]*

---

[1] Net Neutrality has multiple aspects but only the "traffic discrimination or not" aspect is highly controversial. We focus on this aspect in this chapter.

Around the same time,

> *"William L. Smith, chief technology officer for Atlanta-based Bell South Corp., told reporters and analysts that an Internet service provider such as his firm should be able, for example, to charge Yahoo! Inc." for the opportunity to have its search site load faster than that of Google Inc.*
>
> *"Or, Smith said, his company should be allowed to charge a rival voice-over-Internet firm so that its service can operate with the same quality as Bell South's offering." [WSHpost1]*

These remarks sparked furious debate in the network industry and beyond. From the Internet Content Providers (ICPs) and some other Net Neutrality proponents' view, if the incumbent broadband access providers are allowed to treat traffic differently at will, they will become the gatekeepers of the Internet. The Internet may not be neutral to all traffic sources and applications. They argued that this is detrimental to the overall economy of society. These groups started lobbying the U.S. Congress to enact Net Neutrality legislation to prohibit broadband carriers from "making discrimination on the Internet." Interestingly, many consumer rights advocacy groups also take a stand on Net Neutrality. By end of 2006, over one million signatures were delivered to Congress in favor of Net Neutrality [SaveNet]. The heated debate and a number of high-profile Congress hearings pushed Net Neutrality and CoS, one possible way to realize QoS, into the spotlight of public opinion.

While some people may "credit" Mr. Whitacre for starting the Net Neutrality debate, it is not really the case. The "network neutrality" term existed before Mr. Whitacre made his statement. In 2005, Tim Wu, then Associate Professor of Law at University of Virginia Law School (now Professor of Law, University of Columbia), published a widely referenced article "Network Neutrality, Broadband Discrimination" [NNwu], which popularized the term "Network Neutrality." The basic concept can be dated back to the age of the telegram. In 1860, a U.S. federal law subsidizing a coast-to-coast telegraph line stated that:

> *"messages received from any individual, company, or corporation, or from any telegraph lines connecting with this line at either of its termini, shall be impartially transmitted in the order of their reception, excepting that the dispatches of the government shall have priority." [NNwiki]*

Telegram and telephone service providers are considered as "common carriers" under U.S. law. This means that they use certain public assets (for example, right of way or licensed spectrum) and offer their services to the general public. Common carriers are overseen by the U.S. Federal Communications Commission (FCC) to ensure fair pricing and access. Common carriers are explicitly forbidden to give preferential treatment.

The regulatory status of Internet service is a little complicated. Cable modem Internet access has always been categorized under U.S. law as an Information Service, not a telecommunications service. Therefore, it has not been subject to "common carrier" regulations. On the other hand, Internet access across the phone network, including DSL, was for a long time categorized as a telecommunications service, and subject to common carrier regulations. However, on August 5, 2005, the FCC reclassified DSL services as Information Services rather than Telecommunications Services, and replaced common carrier requirements with a set of four less-restrictive Net Neutrality principles [NNfcc] (will be described in the "FCC's Plan" section in this chapter). Note that this shortly preceded Mr. Whitacre and Mr. Smith's remarks. This sparked a debate over whether or not incumbent broadband access providers should be allowed to differentiate between different content providers by offering preferential treatment to higher-paying companies and customers, thus allowing some services to operate faster or more predictably and ultimately become more acceptable to end users. In other words, it is the FCC's reclassification that caused the Net Neutrality debate.

The Interest Conflict between the "Hosts" and the "Parasites"

Whereas many groups are involved in the Net Neutrality debate, the groups that have direct interest conflict are the ICP group and the incumbent NSP group. In this book, the ICP term is extended to include any business that provides services on the Internet. Such services include VoIP, although voice is not traditionally considered as content.

Internet video and some P2P-based file sharing applications push a lot of traffic. They increase NSP's Capital Expenditure (CAPEX) for additional network equipments and Operations Expenditure (OPEX) for managing larger and more complicated networks. VoIP relies on the NSP's network while directly competes with the incumbent NSP's own telephone service. But these ICPs pay just the network access fee. From this perspective, some incumbent broadband access providers feel that the major ICPs are like "parasites," whereas they are like the victimized "hosts." At a time when incumbent broadband access providers are facing record losses of telephone subscribers to mobile operators and VoIP providers, and at the same time facing triple-play competition from the cable operators, they are trying every way to find new revenue sources. In contrast, many ICPs are making big profit. Therefore, the NSPs would like to get "fair" and charge these ICPs a fee for QoS which, in their opinion, is essential to the ICPs' services. This sentiment was exemplified by Verizon Senior Vice President and Deputy General Counsel John Thorne's remark:

> "The network builders are spending a fortune constructing and maintaining the networks that Google intends to ride on with nothing but cheap servers. It is enjoying a free lunch that should, by any rational account, be the lunch of the facilities providers." [WSHpost2]

From the ICPs' perspective, end users have already paid their broadband access fee. ICPs also pay their Internet access fee. Therefore, NSPs have already been compensated for building the networks. They are not entitled to charge another QoS fee. They also argue that at most places in the United States, DSL and cable operators are effectively a duopoly. Some people even argued that the NSPs should thank and pay the ICPs for making the Internet useful, just like cable operators pay content providers today. Furthermore, ICPs argued that if Net Neutrality protection is not enacted, the Internet's freedom could be compromised, limiting consumer choice, economic growth, and technological innovation. Vinton G. Cerf, Vice President and "Chief Internet Evangelist" at Google, said in an interview:

> *"In the Internet world, both ends essentially pay for access to the Internet system, and so the providers of access get compensated by the users at each end. . . . My big concern is that suddenly access providers want to step in the middle and create a toll road to limit customers' ability to get access to services of their choice even though they have paid for access to the network in the first place." [WSHpost2]*

The Net Neutrality Proponent's View

The prominent advocates in the proponent's camp include Vint Cerf, Vice President and Chief Internet Evangelist of Google, Lawrence Lessig, Professor of Law at Stanford University, and Tim Wu, Professor of Law at Columbia University. Although people's opinion on Net Neutrality varies, the key points of the proponents were best expressed by Cerf in his testimony to the U.S. Senate in February 2006 [NNcerf]:

- Most American consumers today have few choices for broadband service. Phone and cable operators together control 98 percent of the broadband market, and only about half of consumers actually have a choice between even two providers. Unfortunately, there appears to be little near-term prospect for meaningful competition from alternative platforms. Consequently, broadband access carriers will have the economic incentive, and capability, to control users' online activity. This is good for the broadband access carriers but bad for the overall economy.

- Monopoly or near monopoly providers should be held to a stricter standard of conduct than when competition and market forces can set prices/standards.

- Network neutrality need not prevent carriers from developing software solutions to remedy end-user concerns such as privacy, security, and quality of service. The issue arises when the network operator decides to place the functionality in the physical or logical layers of the network, rather than in the application layer where they belong. With a few very narrowly tailored exceptions—such as defending against network-level denial of service attacks or router attacks—altering or blocking packets within the network is inconsistent with the end-to-end design principle. The end result is the insertion of a gatekeeper that—even arguably under the best of intentions—disrupts the open, decentralized platform of the Internet.

- Carriers can set market prices for Internet access and be well paid for their investments—as broadband carriers in other countries have successfully done. High-bandwidth Internet applications such as video could be subject to additional customer charges, based on the access speeds required (as opposed to the source, destination, or content of the traffic).

- Broadband access providers have agreed to build robust broadband platforms to support the Internet for reduced regulation at the FCC. Saying that Net Neutrality would remove their incentives to continue investing is turning away from those commitments.

- Facilities-based incumbents are both providers of network access *and* competitors to services that might be offered over that network access (for example, POTS from the incumbent vs. VoIP from Vonage). It would be naïve to assume that the incumbent could fulfill both roles fairly without regulatory guidance.

- Consumers should be able to use the Internet connections that they pay for the way that they want. It should be the users who pick winners and losers in the Internet marketplace, not the carriers.

- While deregulation of our telecommunications system is generally welcome, some limited elements of openness and non-discrimination have long been part of our telecommunications law and must be preserved. Absent real physical-layer competition, a tailored, minimally intrusive, and enforceable Network Neutrality rule is needed.

Note that the Net Neutrality opponents dispute some of the arguments and statistics here. For example, they do not agree that phone and cable operators together control 98 percent of the broadband market. See their views in the next section. This book takes no position on whether these statistics are accurate. We simply present both camps' view for the readers to make their own judgment.

Similarly, in his testimony to the U.S. Senate, Professor Lawrence Lessig argued that allowing incumbent broadband access providers to treat different ICPs' traffic differently based on whether they pay a QoS fee (this is termed "access tiering" by Professor Lessig) would be detrimental to the Internet economy. He proposed that Congress should add a restriction on access tiering to the four policy principles on Internet freedom expressed by former FCC Chairman Michael Powell (which will be discussed later in the "U.S. FCC's Plan" section), and make them law for the FCC to enforce. In the "Network Neutrality, Broadband Discrimination" article, Professor Tim Wu argued that to address the lack of broadband competition, structural separation (will be discussed later) or carrier self-regulation may not be the best alternative, and some legal guidelines may be useful. Wu provided a template. It's quoted below for the user's information.

(a) Broadband Users have the right reasonably to use their Internet connection in ways which are privately beneficial without being publicly detrimental.

Accordingly, Broadband Operators shall impose no restrictions on the use of an Internet connection except as necessary to:

1. Comply with any legal duty created by federal, state or local laws, or as necessary to comply with any executive order, warrant, legal injunction, subpoena, or other duly authorized governmental directive;
2. Prevent physical harm to the local Broadband Network caused by any network attachment or network usage;
3. Prevent Broadband users from interfering with other Broadband or Internet Users' use of their Internet connections, including but not limited to neutral limits on bandwidth usage, limits on mass transmission of unsolicited email, and limits on the distribution of computer viruses, worms, and limits on denial-of service or other attacks on others;
4. Ensure the quality of the Broadband service, by eliminating delay, delay variation or other technical aberrations;
5. Prevent violations of the security of the Broadband network, including all efforts to gain unauthorized access to computers on the Broadband network or Internet;
6. Serve any other purpose specifically authorized by the Federal Communications Commission, based on a weighing of the specific costs and benefit of the restriction.

Note that inside the Net Neutrality camp, different groups have different opinions on the exact meaning of Net Neutrality. Therefore, not every Net Neutrality proponent will agree with every proposed legal clause here.

More information about the proponent's views can be found at the web site of the Save the Internet coalition: http://www.savetheinternet.com/. This web site also contains links to the member organizations that support Net Neutrality.

The Opponent's View

The prominent opponents of Net Neutrality include Scott Cleland, a telecom analyst, and Robert Kahn, co-inventor of TCP/IP (with Vint Cerf) and Chairman, CEO, and President of the Corporation for National Research Initiatives (CNRI, http://www.cnri.reston.va.us/). Some of the key points are summarized below:

- Net is not neutral today. First, Internet traffic is not treated neutrally today; large entities that invest more in infrastructure and pay more, for example, for the premium Internet service provided by Akamai, get better Internet service. Second, net access pricing is not neutral: Internet access price differentiation is the norm. Consumers can choose from a wide variety of Internet price/speed tiers. Prices differ greatly depending on which bundled products or services one buys and for what time period. Third, net usage is not neutral. A small number of users consume most of the Internet's bandwidth because they use highly bandwidth-intensive applications like peer-to-peer video-file-sharing/gaming and video. Five percent of Net users use 51 percent of the bandwidth and 25 percent use 85 percent overall, per Time Warner Cable. Therefore, Net Neutrality, which denies NSPs the right of pricing flexibility, is reverse Robin Hood: Average users must subsidize the heavy users.

- Broadband competition is increasing
  - Choice of broadband providers is expanding rapidly: According to the most recent FCC data: 81 percent of U.S. zip codes offer three or more broadband choices, up from 61 percent in 2003, and 32 percent in 2000, 53 percent of U.S. zip codes offer five or more broadband choices, up from 35 percent in 2003, and 15 percent in 2000. Zip codes with ten or more broadband choices have exploded ninefold since 2000 from 2 percent to 21 percent [FCChss].
  - A good indication of increasing broadband competition is real prices for broadband are falling. Real DSL prices have fallen ~50 percent as speeds have roughly doubled over the last two years; introductory DSL prices have fallen ~70 percent in ~3 years; average monthly DSL prices fell ~15 percent from 2004–2005. Real cable modem prices have fallen ~70 percent as speeds have increased from 1.5 Mbps to 5+ Mbps over the last two years with no price increase. Cable modem prices as part of a bundle have also fallen.
  - The supply of new broadband competitors continues to increase. Wireless broadband is the fastest-growing broadband option: Verizon, Sprint, and AT&T now offer wireless broadband service in most of the country. T-Mobile broadband service offers eight thousand Wi-Fi hotspots with coverage in all fifty states. Several hundred U.S municipalities are in the process of installing citywide Wi-Fi networks.
  - Broadband duopoly allegation is a gross misrepresentation of this dynamic marketplace. Those who allege a telco-cable duopoly egregiously omit the factual context that they both used to be monopolies and that bipartisan competition policy has successfully de-monopolized these markets.

- Market competition will work: Declining costs are lowering the cost of market entry and challenging incumbents. Vibrant innovation rewards competitors with growth from new products, services, and content. Exploding spectrum availability enables more competitors and more competitive applications.

Note that these statistics are different from those presented by the Net Neutrality proponents. The opponents like to point out that the proponents quoted outdated FCC statistics. The proponents countered by saying that there is a loophole in the accounting approach of the new FCC statistics. This book takes no stand on whose statistics are more accurate.

Some other points expressed by Net Neutrality opponents in various circumstances included:

- Net Neutrality is essentially price regulation on the Internet. In a market economy, NSPs should be free to use their assets in an economically rational way. To encourage NSPs to make additional investments in infrastructure, the government shouldn't restrict the NSPs' pricing flexibility.

- Consumers are familiar with other examples of differentiated services with differential pricing, e.g., U.S. mail has multiple classes of services; airlines have first, business, and economy classes. Network services need not be different.

- Net Neutrality is a solution looking for a problem. Problems with differential pricing are all theoretical so far. It's premature to introduce a remedy when the symptom doesn't even exist.

- Differentiated services are technically needed to accommodate voice, video, and other advanced services.

- We already have the FCC that's supposed to be handling communication regulation. We also have another agency, the FTC, that's supposed to be insuring businesses conduct their business in a fair and competitive manner. Therefore, new Net Neutrality legislation is not needed.

More information about the opponent's views can be found at: http://www .netcompetition.org.

The General Public and the Network Industry's View

In the general public, it appears that more people speak out for Net Neutrality than against it. By the end of 2006, over one million signatures were delivered to Congress in favor of Net Neutrality [SaveNet]. As of Jan. 2007, according to a poll by Motley Fool (http://www.fool.com/), 64 percent of the general public favor Net Neutrality legislation, 30 percent oppose, and 7 percent haven't made up their mind [NNpoll]. But more people speaking out for it doesn't necessarily mean that more people actually support it. It could just be the proponent's camp is more vocal than the opponent's camp. In addition, some people might have taken a stand on the issue without fully understanding it. That is, some people might support Net Neutrality just because "keep the Internet neutral" sounds like a moral and appealing cause. The fact that Motley Fool, a financial web site, is doing a poll on Net Neutrality kind of indicated that.

In the networking/Telecom industry, professionals are more divided on Net Neutrality. The primary reason for the higher percentage of reservation may be that these professionals are generally better informed on this issue. They may feel that the NSPs need to invest in infrastructure to enable these premium applications such as voice or video, therefore, they should be allowed to charge a QoS fee to recover their investment. They may realize that it is impractical for networks to become totally neutral. For one thing, virus, Distributed Denial of Service (DDOS) attacks, SPAM, and bandwidth hogs all need to be controlled. Yet another part may be that telecom professionals have been talking about QoS for more than a decade. It's difficult to let it go overnight. But as we will see later, even many network professionals may not be fully informed on Net Neutrality. We said so because first, Net Neutrality doesn't necessary prevent NSPs from recovering their investment in network infrastructure, and second, Net Neutrality is not really about neutrality of the networks, despite what the name conveys.

Clarifying Some Common Misconceptions about Net Neutrality

From the online debates on Net Neutrality, it appears that there are a number of common misconceptions. Clarifying them will help people understand the real issue on Net Neutrality.

- Misconception #1: Incumbent broadband access providers are "common carriers" using public assets (such as right of way or licensed spectrum) on the public's behalf. Therefore, they should not be allowed to discriminate in their marketing of those resources.

Legally, it is a misconception to call cable modem or DSL a "common carrier" service. Cable modem has been an "information service" for a long time. DSL was a "common carrier" service but not anymore. FCC reclassified it as an "information service" in August 2005. Those are legal facts. Unless we want to overturn past legal decisions, Net Neutrality proponents cannot cite the "common carrier" requirement as a basis for Net Neutrality.

- Misconception #2: Without Net Neutrality NSPs may block certain web sites that refuse to pay them a QoS fee. Therefore, Net Neutrality legislation is necessary.

Although some NSPs may appear to have the intention to do that in the past, they no longer plan to do so, at least not in the foreseeable future. AT&T, in particular, stated that Mr. Whitacre was misinterpreted and AT&T never planned to do so on the Internet [WSHpost3]. Plus, the FCC has a clear policy statement against such behavior and has showed a willingness to enforce such policy [FCCmrc]. Therefore, this can't be used as a strong justification for Net Neutrality.

- Misconception #3: Net Neutrality proponents ask for total network neutrality, and NSPs can't even discard or deprioritize virus, SPAM, and DDOS traffic.

This may be an attempt to make Net Neutrality look absurd. Net Neutrality proponents never propose that.

- Misconception #4: The Internet is neutral today. If it stays neutral to all sources, destinations, and applications, startup companies can compete at the same level with established companies. Therefore, Net Neutrality will promote innovation.

First, what is neutral, like what is fair, is hard to define. Arguably, the Internet has not been completely neutral from day one. The Internet's Best-Effort service is more suitable for non-delay/delay-variation-sensitive applications, such as emails, than for delay/delay-variation-sensitive applications such as VoIP. Some applications (for example, those based on TCP) are more cooperative to avoid congestion than other applications (for example, those based on UDP). Therefore, it is a misconception that the Internet is neutral today. Even for the same type of application, some senders can afford Akamai's premium service whereas others can't. With Akamai's content-caching network, its customers' videos can load faster and have better perceived quality than non-Akamai-customers' videos. With Internap's premium Internet service, its customers' web sites may appear more reliable and faster than non-Akamai-customers' sites. Given that no one is proposing to outlaw the services of Akamai, Internap, and the like, Net Neutrality will not enable all companies to compete at the same level. It's a misconception to think otherwise.

- Misconception #5: Premium applications such as video or VoIP need QoS to work. To provide QoS, NSPs must be able to differentiate packets at the needy time (e.g., when there is traffic congestion because of unexpected events), and must also invest in network infrastructure to be able to do so. Therefore they should be allowed to charge a QoS fee to recover the investment. Consequently, Net Neutrality advocators' call for prohibiting NSPs' ability to charge for QoS is unjustifiable.

This is a widely held misconception. Many people cited it to oppose Net Neutrality. The reasons it is a misconception are: First, both consumers and ICPs are already paying for their access. NSPs do not necessarily need to charge a QoS fee to recover their infrastructure investment; second, the mainstream Net Neutrality advocators don't propose prohibiting all companies from charging for QoS. Net Neutrality is very focused on the incumbent broadband access providers, such as phone companies and cable operators. Akamai and Internap are effectively selling QoS today. Nobody has proposed to end that. Net Neutrality advocators didn't even propose prohibiting incumbent broadband access providers from charging for QoS. Instead, it calls for legal scrutiny of the specific methods to charge for QoS, because they believe some methods may be detrimental to the whole economy of the society while other methods may not be. For example, if after upgrading its network to support QoS, an incumbent broadband access provider decides to recover its QoS investment by pricing QoS into its broadband access service, thus raising the broadband access fee, no Net Neutrality proponent has opposed that. Of course, the clarification in this paragraph does not serve to say that Net Neutrality legislation is needed. It only serves to say that this conception cannot be a "killer reason" to throw out Net Neutrality.

- Misconception #6: Net Neutrality is primarily to protect the interest of the broadband users. But broadband access providers will do best if they serve the best interest of their customers, the broadband users, too. Therefore, in pursuit of their best interest, broadband access providers' self-regulation will serve the purpose of any meaningful Net Neutrality proposal. Explicit legislation is not needed. When more and more people realize this, the Net Neutrality issue will fade away.

This is a misconception because some Net Neutrality advocates long recognized that even the monopoly NSPs can have common interest with the broadband users and the content providers, but they still proposed Net Neutrality legislation. As Professor Tim Wu put it in [NNwu]:

"... a platform monopolist has a powerful incentive to be a good steward of the applications sector for its platform.... A monopolist may still want competition in its input markets, to maximize profit in the monopoly market.... But it is easy for a steward to recognize that the platform should support as many applications as possible. The more difficult challenge has always been the dynamic aspect: recognizing that serving a tangible goal—like controlling bandwidth usage—may affect the intangible status of the Internet as an application development platform. Some of the restrictions, such as those on running various types of server, are applications that are now likely to be used by only a small minority of broadband users. Their sacrifice may appear like a good cost saving measure.

"More generally, the idea that discrimination may not always be rational is a well-understood phenomenon.... Broadband operators may simply disfavor certain uses of their network for irrational reasons, such as hypothetic security concerns or exaggerated fears of legal liability. Additionally, a restriction may become obsolete: adopted at a certain time for a certain reason that no long matters. Practical experience suggests that such things happen."

An example of the above is: While most people accept that residential broadband service cannot be used for commercial purpose, Cox, one of the major cable operators in the United States, once barred home users from using Virtual Private Network (VPN) service to connect to their employer's corporate network because that was considered commercial usage. Today, this would look absurd because VPN is allowed by virtually all NSPs. In 2003, Cox rescinded this restriction.

There were also other examples. At one point, connecting a computer to a telephone line or connecting multiple computers to a broadband connection was considered unacceptable user behavior. The "justification" for the former is, the phone line was provided for phone service, not dialup data service. The "justification" for the latter is, the broadband connection is for one computer only.

Because of incidences like these, Professor Wu argued that:

"For these reasons, anti-discrimination regulation or the threat thereof can also serve a useful educational function. It can force broadband operators to consider whether their restrictions are in their long-term best interests."

In other words, the Net Neutrality issue will not fade away because people recognize the goodwill of the broadband access providers.

- Misconception #7: FCC already has the tools needed for Net Neutrality. New legislation is not needed.

This is a misconception because there is no FCC policy regarding traffic discrimination. FCC does have policies guarding other aspects of Net Neutrality outside of QoS. For more information, refer to the "U.S. FCC's Plan" section below.

From the above we can see that both camps have some misconceptions about the other camp. For people in one camp, misconceptions cause the other camp to appear unreasonable or even absurd. This in turn causes many people to take a strong and even emotional stand on Net Neutrality without fully understanding all the facets of it. For example, if a person thought that a NSP was going to block Google tomorrow unless Google paid a QoS fee today, that person will likely feel that the NSP is acting unreasonably. The person may therefore favor Net Neutrality legislation. The person will not realize that the Net Neutrality controversy is more complicated than that.

Despite the author's effort to stay objective on Net Neutrality, it is expected that not all people will agree with the clarifications here. We would like to point out that this does not matter to the theme of this book, which is focused on QoS. This is because, while there are debates on some complicated aspects of Net Neutrality, there should be no debate on the point that Net Neutrality is concerned with: whether NSPs should be allowed to treat different companies/people's traffic differently based on how much they pay. In QoS terms, this concerns whether NSPs should be allowed to do CoS-based pricing and prioritize traffic accordingly. The traditional QoS business model takes it for granted that NSPs can do so. For the QoS discussion in this book, the purpose of this chapter is to say that this issue is at the center of the Net Neutrality debate and cannot be taken for granted. Therefore, the controversy on other aspects of Net Neutrality does not matter.

The Gist of Net Neutrality

While "the Internet should be neutral to all sources, all destinations, and all applications" may sound like a noble cause to fight for, Net Neutrality is not about whether the Internet should be neutral or fair to all. It is about how to address the lack of meaningful broadband access competition in the United States. According to the testimony of Vint Cerf, *"Were there sufficient competition among and between various broadband networks, Google's concerns about the future of the Internet would largely be allayed,"* and *"The best long-term answer to this problem is significantly more broadband competition." [NNcerf]*

It is only because *"the prospects for such 'intermodal' competition remain dim for the foreseeable future, Congress should ensure that the FCC has all the tools it needs to maximize the chances for long-term success in this area."*

Similarly, Professor Lawrence Lessig stated, *"It was the assumption of many (including me) that competition in broadband access would prevent any compromise in end-to-end neutrality". [NNlessig]*

Similarly, one of the prominent opponents of Net Neutrality, Robert Hahn expressed the opinion that competition is the solution, not government regulation.

"We (Robert Hahn and Scott Wallsten) argue that mandating net neutrality would be likely to reduce economic welfare. Instead, the government should

focus on creating competition in the broadband market by liberalizing more spectrum and reducing entry barriers created by certain local regulations." *[NNhahn]*

The main difference between the two camps is whether there is enough broadband competition, and whether there will be enough competition in the near future, such that market force can take care of the issue without government regulation. This is the gist of Net Neutrality.

Note that Net Neutrality is primarily focused on residential broadband networks. Private corporate networks and NSPs' special purpose networks are less of a concern. However, Cerf argued that:

"Some carriers are also seeking permission to create two separate IP networks: one for the public Internet and one for a privately-managed, proprietary service. Allowing segmentation of the broadband networks into capacious "broadest-band" toll lanes for some, and narrow dirt access roads for the rest, is contrary to the design and spirit behind the Internet, as well as our national competitive interests. And by definition, favoring some disfavors others. In an environment where consumers already have little to no choice of broadband providers, the end result is a cramped version of the robust and open environment we all take for granted today. Prioritization inevitably becomes a zero-sum game." [NNcerf]

In other words, there is also some uncertainty regarding charging for QoS in NSPs' special-purpose networks (for example, corporate service networks), although not as much.

Can Structural Separation Be an Alternative to Net Neutrality?

In the Net Neutrality debate, there is a third camp other than the proponent and the opponent. This third camp agreed with the proponents that there is a lack of broadband competition and the issue needs to be addressed. But they also agreed with the opponents that Net Neutrality law can be vague and may be difficult to enforce. It is possible that such law will end up with a lot of lawsuits that only slow down network development. Given that lack of competition is the key issue, they argued that the remedy should be increasing competition. Their proposed way to do so is through structural separation.

Structural separation is also known as Horizontal Segmentation, Open Access, or Local Loop Unbundling (LLU). The basic idea is to segment the currently vertically integrated incumbent broadband access providers into two separate business entities: a broadband access infrastructure provider (that is, wholesaler) and a service/content provider (that is, retailer). The wholesaler must provide packet transport services to the incumbent retailer and all other third-party retailers in the same terms. The end users are the customers of the retailers. The effect of structural separation is similar to "unbundling" that is adopted by the European regulators.

The arguments for structural separation are as follows. First, because the wholesaler does not have its own retail services, it will stay neutral to all retailers. It may be required by law to do so as well. Second, structural separation would enable new parties to enter the service retail market, and increase competition there. The increased competition would benefit the consumers. The proponents of structural separation generally like to point out that structural separation (that is, LLU) works well in the United Kingdom and France.

The arguments against structural separation are: First, as a "utility" type of company whose price is regulated by the government, infrastructure providers will not have incentive to upgrade their networks. After all, whether the applications riding on the network work or not has little economic relevance to them. This will result in a slow network for everybody. Second, some people argued that sophisticated applications such as video or other real-time applications require special support from the network (of course, this is itself a contentious point), and structural separation would prevent that from happening. Professor Tim Wu argued that this would amount to discrimination against real-time applications.

The United States used to have unbundling requirements for the Incumbent Local Exchange Carriers (ILECs), which became the incumbent DSL providers of today. It's called Unbundled Network Equipment-Platform (UNE-P), through which another company can make use of the telephone local loops owned by the ILECs to provide DSL service. Combined with abundance of venture capital in the 1999 to 2000 Internet bubble time, this led to a plethora of DSL service companies and in turn, aggressive price competition. Most of these companies went out of business when the bubble burst and the capital infusion stopped. Fair or not, this is regarded by some people as evidence that unbundling did not work. In 2003, FCC removed UNE-P as a generic requirement for the ILECs to encourage them to invest in broadband infrastructure. Structural separation would represent a reverse of that direction. Given how daunting the task is, there is currently no strong lobby to the U.S. government to reinstitute unbundling. Structural separation is more or less an undercurrent at this moment. However, a few Wall Street financial analysts argued that structural separation would increase the market value of the incumbent broadband access providers. The proponents of structural separation hope that financial market pressure and the development of municipal networks that adopt the open access model would eventually lead to structural separation. It should be pointed out that incumbent broadband access providers in the United States will likely strongly oppose structural separation. They may even accept Net Neutrality legislation to avoid structural separation, given that both AT&T and Verizon, the two largest incumbent broadband access providers in the United States, have been trying very hard to acquire other NSPs to create a large, all-encompassing service provider.

There is no clear indication where the Net Neutrality debate will lead to. Therefore there is much uncertainty related to the commercialization of QoS as per the traditional QoS business model.

THE CONSEQUENCES OF THE NET NEUTRALITY DEBATE

The immediate consequence of the Net Neutrality controversy is uncertain government regulation. Both camps had moments when they thought that they were close to victory. But as of early 2007, it was still not clear whether the U.S. government will eventually enact a Net Neutrality law.

Government Regulation Uncertainty

The two government organizations that are closely involved in the Net Neutrality controversy are the U.S. Congress and the FCC. Their plans on Net Neutrality are briefly reviewed below.

U.S. Congress's plan

Both the U.S. Senate and the House of Representatives are lobbied by both sides in the Net Neutrality debate. Both chambers have draft bills that favored or disfavored Net Neutrality. Below is a list of draft bills and their stand on Net Neutrality. So far none of them has become law.

- **HR5252**, "Communication Opportunity, Promotion and Enhancement Act of 2006": It disfavored Net Neutrality by excluding the legal clauses desired by the Net Neutrality proponents

- **HR5273**, "Network Neutrality Act of 2006": It favored Net Neutrality

- **HR5417**, "Internet Freedom and Nondiscrimination Act of 2006": It favored Net Neutrality

- **S1504**, "Broadband Investment and Consumer Choice Act": It favored Net Neutrality

- **S2360**, "Internet Non-Discrimination Act of 2006": It favored Net Neutrality

- **S2686**, "Communications, Consumer's Choice, and Broadband Deployment Act of 2006": It disfavored Net Neutrality

- **S2917**, "Internet Freedom Preservation Act (of 2006)": It favored Net Neutrality

- **S215**, "Internet Freedom Preservation Act (of 2007)": It favored Net Neutrality

S215 is worth expanding on a bit because it is the latest one, and its introducers include some of the most powerful politicians in the United States, including Senator Barack Obama and Senator Hillary Clinton. A few key passages related to traffic discrimination and related charging are enclosed below. The complete bill can be found in Appendix B.

(a) Duty of Broadband Service Providers—With respect to any broadband service offered to the public, each broadband service provider shall—

(1) not block, interfere with, discriminate against, impair, or degrade the ability of any person to use a broadband service to access, use, send, post, receive, or offer any lawful content, application, or service made available via the Internet;

(2) enable any content, application, or service made available via the Internet to be offered, provided, or posted on a basis that—

(A) is reasonable and nondiscriminatory, including with respect to quality of service, access, speed, and bandwidth;

(B) is at least equivalent to the access, speed, quality of service, and bandwidth that such broadband service provider offers to affiliated content, applications, or services made available via the public Internet into the network of such broadband service provider; and

(C) does not impose a charge on the basis of the type of content, applications, or services made available via the Internet into the network of such broadband service provider;

(3) only prioritize content, applications, or services accessed by a user that is made available via the Internet within the network of such broadband service provider based on the type of content, applications, or services and the level of service purchased by the user, without charge for such prioritization; and

(b) Certain Management and Business-Related Practices—Nothing in this section shall be construed to prohibit a broadband service provider from engaging in any activity, provided that such activity is not inconsistent with the requirements of subsection (a), including—

(1) protecting the security of a user's computer on the network of such broadband service provider, or managing such network in a manner that does not distinguish based on the source or ownership of content, application, or service;

(2) offering directly to each user broadband service that does not distinguish based on the source or ownership of content, application, or service, at different prices based on defined levels of bandwidth or the actual quantity of data flow over a user's connection;

If enacted as law, S215 would mandate that:

- NSPs can differentiate traffic based on application, but can't differentiate traffic based on end users or application service providers;

- NSPs can differentiate traffic for network management purpose;

- NSPs cannot charge end users or application service providers for the differentiation.

This is consistent with our clarifications in the previous section.

In summary, from these draft bills it's clear that the U.S. Congress is also divided on Net Neutrality. When the Congress was controlled by the Republicans, the anti-regulation view was more popular. HR5252, a draft bill supported by the Net Neutrality opponents, almost became law when the Republicans were in control of the Congress. Now that both the Senate and the House are controlled by the Democrats, the possibility of Net Neutrality legislation has increased.[2]

U.S. FCC's plan

Since the telecommunication deregulation in 2003, the FCC has been trying to design policies that protect consumer rights while giving the NSPs the incentive to upgrade their networks. In February 2004, then FCC Chairman Michael Powell outlined four principles that he promised would guide FCC policy:

Consumers are entitled to:

1. Freedom to Access Content.
2. Freedom to Use Applications.
3. Freedom to Attach Personal Devices.
4. Freedom to Obtain Service Plan Information.

In August 2005, under the leadership of new chairman Kevin Martin, the FCC issued a policy statement that clarified these four principles [NNfcc]:

1. Consumers are entitled to access the lawful Internet content of their choice;
2. Consumers are entitled to run applications and services of their choice, subject to the needs of law enforcement;
3. Consumers are entitled to connect their choice of legal devices that do not harm the network; and
4. Consumers are entitled to competition among network providers, application and service providers, and content providers.

The FCC further clarified that:

"Although the Commission did not adopt rules in this regard, it will incorporate these principles into its ongoing policymaking activities. All of these principles are subject to reasonable network management."

This is generally viewed as FCC's statement that it already has the needed tools to protect consumer rights and to enforce Net Neutrality. FCC Chairman Kevin Martin was quoted in an interview stating, "I don't think that there is evidence of the kind of activity of blocking consumers' access to the Internet that would justify us adopting new rules at this stage."

---

[2] For people unfamiliar with the political system in the United States, it may be helpful to know that historically Republicans are more sympathetic to big business and less keen on government regulations, while Democrats are more sympathetic to small business and individuals, and are more willing to regulate big business.

The Net Neutrality proponents pointed out that none of the principles above deals with traffic discrimination, that is, CoS. They therefore argued that Net Neutrality legislation by the Congress is needed. One of Lawrence Lessig's proposals is to add a restriction on "access tiering" (that is, forbidding incumbent broadband access providers from providing multiple CoS's to content providers and prioritizing their traffic accordingly) to the existing FCC principles, and make them explicit rules for FCC (that is, not just policies).

THE IMPACT OF NET NEUTRALITY ON QOS

From a QoS perspective, Net Neutrality is regarding whether NSPs should be allowed to do CoS-based pricing and differentiate traffic accordingly. The traditional QoS business model takes it for granted that NSPs can do so. Net Neutrality challenges this fundamental assumption. Therefore, it casts serious doubt on the traditional QoS business model. If Net Neutrality is adopted as law, the traditional QoS business model will be invalidated. Even if Net Neutrality is just unsettled, it poses significant uncertainty for the traditional QoS business model. This will discourage NSPs from adopting the traditional QoS business model, for fear of reigniting the debate. As a matter of fact, a likely outcome of the Net Neutrality controversy is the status quo is maintained. That is, there will be no Net Neutrality legislation, but there will be no selling of CoS or traffic differentiation on the Internet either. The plan that AT&T submitted to the FCC in December 2006 when it sought approval for its merger with Bell South offered early evidence of that.

1. Effective on the Merger Closing Date, and continuing for 30 months thereafter, AT&T/Bell South will conduct business in a manner that comports with the principles set forth in the Commission's Policy Statement, issued September 23, 2005 (FCC 05–151).

2. AT&T/Bell South also commits that it will maintain a neutral network and neutral routing in its wireline broadband Internet access service. *This commitment shall be satisfied by AT&T/Bell South's agreement not to provide or to sell to Internet content, application, or service providers, including those affiliated with AT&T/Bell South, any service that privileges, degrades or prioritizes any packet transmitted over AT&T/Bell South's wireline broadband Internet access service based on its source, ownership or destination.* This commitment shall apply to AT&T/Bell South's wireline broadband Internet access service from the network side of the customer premise equipment up to and including the Internet Exchange Point closest to the customer's premise, defined as the point of interconnection that is logically, temporally or physically closest to the customer's premise where public or private Internet backbone networks freely exchange Internet packets.

This commitment does not apply to AT&T/Bell South's enterprise managed IP services, defined as services available only to enterprise customers that are separate services from, and can be purchased without, AT&T/Bell South's wireline broadband Internet access service, including, but not limited to, virtual private network (VPN) services provided to enterprise customers. *This commitment also does not apply to AT&T/Bell South's Internet Protocol television (IPTV) service.* These exclusions shall not result in the privileging, degradation, or prioritization of packets transmitted or received by AT&T/Bell South's non-enterprise customers' wireline broadband Internet access service from the network side of the customer premise equipment up to and including the Internet Exchange Point closest to the customer's premise, as defined above.

This commitment shall sunset on the earlier of (1) two years from the Merger Closing Date, or (2) the effective date of any legislation enacted by Congress subsequent to the Merger Closing Date that substantially addresses "network neutrality" obligations of broadband Internet access providers, including, but not limited to, any legislation that substantially addresses the privileging, degradation, or prioritization of broadband Internet access traffic.

This means that AT&T will not be charging any QoS fee and differentiating traffic accordingly in at least 2007 and 2008. Furthermore, although this is just AT&T's policy, it will likely induce other NSPs to follow suit.

In summary, Net Neutrality casts doubt in the fundamental assumption of the traditional QoS business model, that NSPs can sell CoS's and differentiate traffic accordingly. The regulatory challenges that Net Neutrality introduced may delay the roll-out of the traditional QoS business model significantly.

While some people may interpret Net Neutrality's effect on QoS as negative, it is not necessarily the case. Net Neutrality exposes many issues of the traditional QoS business model. It helps people to realize that the CoS-centric approach may not work. This can help QoS to take a new approach. QoS need not necessarily take the traditional approach. More specifically, QoS does not necessarily have to be sold explicitly as a CoS higher than Best Effort. Therefore, in retrospect, it is possible the Net Neutrality debate became the turning point where QoS starts to become reality. Of course, this new QoS will be different from the previously narrowly defined CoS-centric QoS. This will be further explored in future chapters.

Although the regulation aspect of Net Neutrality is largely a U.S. issue, its impact on the traditional QoS model has a global effect. This is because the traditional QoS model relies on differentiation, and differentiation is only meaningful if it has an effect on the user applications. That is, local differentiation may not be enough. Because the U.S. networks are an important part of the Internet, if there is no differentiation in it, the differentiation story will not be convincing in other countries.

REGULATORY ENVIRONMENT IN OTHER COUNTRIES

Net Neutrality as a network philosophy has global influence. However, its regulatory implication is more limited to the United States. In this section, we briefly review the regulatory environment related to traffic prioritization in other countries.

As we previously stated, in the United States, the desire to restrict NSPs from prioritizing traffic based on whether a QoS fee is paid comes from the concern of lack of broadband access competition. In other words, Net Neutrality legislation is a proposed solution to the (perceived) problem of lack of broadband access competition. In other countries, the competition statuses can be different. When there is a similar lack of competition, the preferred solution can be different, too. At a high level, it may be fair to say that regulatory bodies in other countries are more tolerant of traffic prioritization, but have not reached any definitive conclusion.

In Europe, there was a lack of broadband access competition, because the last mile infrastructure is largely controlled by the incumbent NSPs. For example, in 2004, the Office of Communications (Ofcom), the regulatory body of the United Kingdom, identified the following challenges for the communications market [Ofcom]:

1. The telecoms sector is changing rapidly as it moves from historical business models based on the delivery of voice calls over switched-circuit networks to business models based on the delivery of data over internet protocol networks.

2. These changes bring uncertainty as well as opportunity, particularly for investors; yet companies have a limited opportunity in time to make the significant, long-term commercial decisions required if they are to remain competitive in the future.

3. The UK telecoms market offers choice and value to the end user in a number of areas, yet despite twenty years of regulatory intervention, competition in fixed line telecoms remains fragile. Additionally, many of the advantages upon which competitors have based their businesses are being eroded, not least by the transition to next generation networks.

4. Consumers' behaviour is changing as new technologies penetrate the mass-market. However, with growth in choice and innovation has come an increase in the potential for confusion, as consumers seek to navigate increasingly complex competitive retail markets.

In seeking to address these challenges, Ofcom identified two key problems to be solved:

- First, an unstable market structure in fixed telecoms, dominated by BT and with alternative providers that are, in the main, fragmented and of limited scale.

- Second, the continuance of a complex regulatory mesh, devised over twenty years of regulation and in many areas dependent upon intrusive micromanagement to achieve its purposes, yet which, in aggregate, has failed effectively to address the core issue of BT's control of the UK-wide access network.

Ofcom presented three options to address these issues:

- Option 1: Full deregulation. Removing the existing mesh of regulation entirely and relying instead on ex post competition law to resolve complaints would significantly reduce intervention in fixed-line markets. However, given BT's continued market power, this would be unlikely to encourage the growth of greater competition and as such would not serve the best interests of the consumer.

- Option 2: Enterprise Act investigation. Ofcom could investigate the market under the Enterprise Act 2002, with the potential for a subsequent referral to the Competition Commission.

- Option 3: BT to deliver real equality of access. Ofcom could require BT to allow its competitors to gain genuinely equal access to its networks. This option would also require BT to commit to behavioural and organisational changes to ensure that its competitors benefited from access to products and processes which were truly equivalent to those offered to BT's own retail businesses.

Option 1 is basically to give up government regulations altogether and rely on the market to regulate itself. Option 2 is to investigate whether BT violates the anti-monopoly stature in the Enterprise Act 2002 and act accordingly. Option 3 is to use structural separation (also called functional separation in Europe, or Local Loop Unbundling, LLU) as a remedy. Ofcom and BT eventually took the structural separation approach. BT separated out the broadband access infrastructure entity as BT Open Reach. All broadband access service retailers, including BT's or a third party's, can use BT Open Reach's infrastructure. According to Viviane Reding, European Commissioner for Information Society and Media,

> *"In the UK, functional separation was successful and unleashed a new wave of investments in the British telecoms industry when the former monopoly BT decided to separate out its network from its services. When it began in 2006, only 200,000 unbundled access lines existed. Since then it has grown to 2 million. Broadband penetration in the UK has at least doubled as a result of this regulatory intervention, while investment was encouraged by this move of the British regulator." [Reding]*

Because of LLU, in the March 2007 debate on Net Neutrality in the United Kingdom, former trade minister Alun Michael stated that preemptive technical legislation like Net Neutrality is ". . . unattractive and impractical"[3] [NNregist]. However, this doesn't mean that Net Neutrality is dead in the United Kingdom. The same report stated that Douglas Scott, Ofcom's Director of Policy, considered *"the European framework permits ISP to prioritize packets by application, which the UK regulator regards as fine."* A grey area, he suggested, *"was when*

---

[3] This debate was sponsored by AT&T. Google considered it biased and didn't participate in this debate.

an ISP offered MySpace a preferential Quality of Service deal, for a fee. Should the regulator constrain the fee?" In other words, whether NSPs can charge a QoS fee is still not clear.

This somewhat self-contradicting view is not specific to U.K. regulators. The 2007 European Commission report "Impact Assessment for the Regulatory Framework for Electronic Communications" [NNeu], first stated that:

> *"In the context of the EU regulatory framework and i2010 Initiative, the debate on net neutrality and freedoms translates into the general concern that the potential of the Internet would be threatened if network or services providers and not users were to decide which content, services, and applications can respectively be accessed or distributed and run.*
>
> *In the US discussion, much of the advocacy to legislatively mandated network neutrality is based on the assumption that differing charges to suppliers of content to the Internet for correspondingly differing speeds of delivery are inherently discriminatory.*
>
> *However, product differentiation is generally considered to be beneficial for the market (particularly in industries with large fixed and sunk costs) so long as users have choice to access the transmission capabilities and the services they want. Allowing broadband operators to differentiate their products may make market entry of content providers more likely, thereby leading to a less concentrated industry structure and more consumer choice.*
>
> *Consequently, the current EU rules allow operators to offer different services to different customer groups (and price such services accordingly), but do not allow those who are in a dominant position to discriminate in an anti-competitive manner between customers in similar circumstances."*
> *[NN-EU]*

While this may appear clear that traffic prioritization is considered fine by European regulators, just a few paragraphs later, the report stated that:

> *"As for 'net neutrality,' the problem also remains that the current regulatory framework does not provide NRAs with the means to intervene were the quality of service for transmission in an IP-based communications environment to be degraded to unacceptably low levels, thereby frustrating the delivery of services from third parties. In such an event, end-users' connectivity to services provided on the internet (TV, telephony, Internet, etc.) could be at risk. The impact of prioritisation or of systematic degradation of connectivity could be larger on services needing real-time communications (e.g. IPTV, VoIP, in which latency is critical) and ultimately affect end-user choice."*

In other words, there is concern that prioritization may lead to unacceptable quality for some users' real-time applications. It is expected that regulators will intervene in that case. So Net Neutrality opponents cannot say "case closed" here, because an ICP can always claim that its real-time services were affected by the NSPs' prioritization. To make the regulators' position even more difficult to understand, Viviane Reding, European Commissioner for Information Society and Media, stated that:

> *"I firmly believe in Net Neutrality. I firmly believe in the principle of access for all. The Commission does not want to see a two-speed Internet where the rich benefit and the poor suffer." [Reding]*

Now let's look at the situation in East Asian countries like Japan, South Korea, and China. In Japan, Nippon Telegraph and Telephone (NTT) operates a service called Flet's Square over its FTTH high-speed Internet connections that serves video on demand at speeds and levels of service higher than generic Internet traffic. So traffic prioritization is apparently allowed. (But of course, the practical version of Net Neutrality does not prohibit all traffic prioritization either.) The author was not able to find out whether there is a QoS fee associated with the traffic prioritization.

In 2006, South Korea banned any foreign-based VoIP companies "not in compliance with Korea's Telecommunications Business Act" from reaching customers using the Internet. Consequently, some of the most popular U.S. VoIP companies like Vonage, AT&T, and Lingo were blocked (for at least a period of time) [VonageF]. The situation in China is similar. NSPs can block VoIP providers' calls, although they may or may not enforce that. If NSPs can block VoIP calls outright, then it is fairly safe to say that they can prioritize traffic as they see fit.

In summary, traffic prioritization is considered acceptable by government regulators in Europe and East Asia, although in Europe there is still concern that such prioritization may have a negative effect on real-time traffic that is not prioritized. However, whether such traffic prioritization will be associated with a QoS fee in Europe or East Asia is less clear. Because differentiation is only meaningful if it has an effect on the user applications, local differentiation may not be enough. Because the U.S. networks are an important part of the Internet, as long as Net Neutrality remains unsettled in the United States, it may be difficult for NSPs in other countries to charge for Internet QoS, even if they want to.

SUMMARY

The key points of this chapter are:

- From a QoS perspective, Net Neutrality is a debate regarding whether incumbent broadband access providers should be allowed to treat different companies/people's traffic differently based on whether a QoS fee is paid. Net Neutrality proponents argued that such differential treatment will lead to "traffic discrimination."

- From the Net Neutrality proponents' perspective, there is a lack of broadband access competition in the United States. Therefore, Net Neutrality legislation is needed to prevent the incumbent broadband access providers from exerting too much control on the Internet.

- From the Net Neutrality opponents' perspective, the Internet is not totally neutral in the past and needs not be totally neutral in the future. There will be more and more broadband competition in the future. The market can take care of itself. Therefore, government regulation like Net Neutrality legislation is not needed.

- Despite the name, Net Neutrality is more about how to address the perceived lack of broadband access competition than about ensuring neutrality/fairness of the Internet.

- The debate on Net Neutrality is unsettled. It is not clear whether Net Neutrality law will be enacted in the United States.

- Structural separation is to separate the incumbent broadband access providers into two separate entities: a last-mile infrastructure provider and a service/content provider. Structural separation is perceived by some people as an alternative to Net Neutrality legislation. There are arguments for and against structural separation.

Net Neutrality challenges the fundamental assumption of the traditional QoS business model, in which it is taken for granted that NSPs can charge for QoS. If Net Neutrality is adopted as law, the traditional QoS business model will be invalidated. Furthermore, even if Net Neutrality is just undecided, it poses significant uncertainty for the traditional QoS business model. This will discourage NSPs from adopting the traditional QoS business model. But this is not necessarily a bad thing for QoS. Net Neutrality may help people to realize that the CoS-centric approach may not work. This can help QoS to take a new approach. That is a topic for discussion in Part 3.

Please voice your opinion about this chapter at http://groups.google.com/group/qos-challenges.

Technical Challenges

8

This chapter will discuss the technical challenges of the traditional QoS solution. This solution is based primarily on Class of Service (CoS), as presented in Chapter 4.

INTEGRATION CHALLENGE

Traditionally, QoS is defined as using Diffserv to create service differentiation. Therefore, most QoS publications focus on traffic management.

But traffic management is just one specific traffic control scheme. Other traffic control schemes can also impact the end users' QoS perception. These traffic control schemes include routing, Traffic Engineering (TE), Content Delivery Network (CDN), etc. In many cases, they can even have a bigger impact on users' QoS perception than traffic management. If traffic management is not integrated with other traffic control schemes, certain challenges will result and the service providers' capability to provide QoS will be limited.

This section discusses various traffic control schemes and their effect on end users' QoS perception. We also point out the challenges that will result if these schemes are not used in a coordinated way.

Traffic Management

Traffic management mechanisms include classification, marking, policing and re-marking, shaping, queueing, and buffer management and packet dispatching. Classification is to separate packets into different categories. Marking involves marking the designated category on each packet based on classification results for subsequent usage. Policing ensures that users only get the service they are entitled to and do not overuse network resources. This could result in re-marking of the user packets. Shaping is making traffic conform to a certain rate. Queueing is deciding the appropriate queue for each packet; buffer management is deciding whether to actually put the packet into the queue. The common buffer

113

management mechanisms are Random Early Detection ([RED]) and Weighted Random Early Detection ([WRED]). They may decide to drop a packet before a queue becomes full, to prevent the queue from overflowing, which will cause bigger damage. So RED and WRED are preventive measures for good TCP performance. Dispatching is to decide which queue at a port will get the opportunity to transmit packets. The common dispatching mechanisms are Strict Priority Queueing (SPQ) and Weighted Fair Queueing (WFQ). With SPQ, queues are assigned a priority order, and they are served strictly according to that order. A low-priority queue may be starved if there are always packets in higher-priority queues. With WFQ, the queues are each assigned a weight, and they share the bandwidth of the port according to that weight. Today many vendors' devices support the mixing of SPQ and WFQ at a port. For example, of the eight output queues associated with a port, one queue may be given highest priority while the remaining queues share the remaining bandwidth. Usually, it is up to the network operators to make sure that the SPQs only carry an appropriate amount of traffic so that lower-priority queues won't be starved.

Together, these traffic management mechanisms can be used to create Diffserv Per Hop Behavior (PHB) at each network device, which will result in the creation of multiple traffic classes in the network. Chapter 4 described such an approach. To use an analogy, Diffserv traffic management is about creating one or multiple express lanes on an information highway. Each express lane would represent a traffic class. If the highway is usually congested, and the amount of high-priority traffic is limited, then this scheme can be useful. However, in the traditional QoS wisdom, these two assumptions are not verified. This is a critical problem and will be further discussed in this chapter.

Routing

Routing is deciding the routes that packets travel in the network. It can be further divided into inter-domain routing and intra-domain routing, controlled by BGP and IGP, respectively.

Inter-domain routing

The Internet consists of many domains. Each domain is called an Autonomous System (AS) and is managed independently. An ISP may own one or multiple ASs. For example, Global Crossing's AS# is 3549, and UUNET (now Verizon Business)'s AS# is 701.

Inter-domain routing is the top-level routing. It determines the AS path each packet will travel through to its destination. Consequently, it has the largest impact on traffic distribution on the Internet. Inter-domain routing policy changes usually have a dramatic effect. They will affect traffic distribution not only inside one's own domain but also in other domains. If used properly, inter-domain routing can be an effective tool for handling congestion. But it must be used with discretion.

Intra-domain routing

Each domain has many routers. Some routers are connected to other domains. They are called border routers. Others are not connected to other domains. They are called interior routers. For a packet that enters a domain, intra-domain routing will determine the route via which the packet will travel through to the border router connected to the next domain.

Intra-domain routing can be used to handle traffic congestion inside a domain, for example, by moving traffic around inside the network. This generally won't affect traffic distribution in other domains. Therefore, intra-domain routing is usually used before inter-domain routing is attempted.

Constraint-based routing

Constraint-based routing refers to a category of routing in which QoS requirements (also called constraints) as well as distance are used as criteria for selecting the optimal path [Wang1][Ma]. For example, each link may be associated with certain delay, delay variation, and packet loss ratio characteristics. In selecting the optimal path between two points, only the paths that meet the constraints can be considered. It can be used with inter-domain routing or intra-domain routing.

While being a hot research topic in the 1990s, constraint-based routing has never been standardized or deployed. First, it is very complicated. In order to do constraint-based routing, each link has to be associated with some delay, delay variation, and packet loss ratio constraint. This adds significant overhead and complexity to the routing protocols. Besides, the path computation algorithm also becomes much more computation intensive, which leads to slower convergence. Second, it is not considered necessary. Common inter-domain and intra-domain routing seem to meet the need of today's real-time applications on the Internet. Again, because QoS needs to be end-to-end, just one NSP doing constraint-based routing is not useful. Consequently, no NSP bothers to be the first deployer.

Constraint-based routing is briefly reviewed in this section for completeness. It will not be further discussed in this book.

Traffic Engineering and MPLS

Traffic Engineering (TE) involves optimizing traffic distribution in one's network [RFC3272]. This can be done with a Diffserv-agnostic manner or a Diffserv-aware manner.

Diffserv-agnostic TE

Common TE can be done by tuning one's inter-domain routing (for example, BGP) policy, or intra-domain routing (for example, OSPF) metrics, or using MPLS. As mentioned previously, changing inter-domain routing policy can dramatically effect traffic distribution not only in one's own domain but also in neighboring domains. Therefore, the inter-domain routing policy is usually carefully designed, and once set, rarely changed unless triggered by commercial reasons, for example, peering or transit contract change. From this perspective, some people may not consider inter-domain routing as a TE mechanism at all. Tuning IGP metrics can change traffic distribution in one's network. Before the coming of Multi-Protocol Label Switching (MPLS) [RFC3031], it is the main TE mechanism. However, because changing the IGP metric of one link can affect all the paths going through that link, without proper simulation tools, tuning IGP metrics to change traffic distribution can have side effects. However, in recent years, some tools emerged [Cariden] and this process has become more scientific.

MPLS was invented as a fast packet-forwarding mechanism, but later TE and other capabilities such as VPN were added. MPLS traffic engineering allows network operators to assign a bandwidth value to each link. The assigned bandwidth usually equals to the physical bandwidth of the link. MPLS traffic engineering also allows the assignment of a bandwidth value to each MPLS tunnel. This value usually equals the amount of traffic carried by the tunnel, as measured. Because the assigned bandwidth values of the links and the Label Switched Paths (LSPs) are signaled by a protocol called Resource Reservation Protocol with Traffic Engineering Extension (RSVP-TE) [RFC3209], these bandwidth values are usually called RSVP bandwidth of the links and the tunnels. When placing the tunnel onto the network, MPLS traffic engineering will ensure that on any link, the bandwidth sum of all tunnels going through that link is less than the bandwidth of the link. Conceptually, this ensures that every link has more bandwidth than needed and thus prevents congestion.

The advantage of MPLS-TE is that traffic distribution can be changed in a very controlled manner, especially with simulation tools such as [WANDL] and [Cariden]. But note that with MPLS and RSVP-TE, each link has to be associated with a bandwidth value. This increases the complexity of the routing protocols. LSPs for TE purposes have to be signaled with an associated bandwidth value, too. MPLS effectively adds a layer into the network and introduces a large amount of additional complexity. If MPLS is also used for other purposes, e.g., Virtual Private Networks (VPNs) [RFC2547], then this may be justified. Otherwise, IGP traffic tuning with a proper simulation tool can be a better alternative because it requires less complexity.

According to WANDL and Cariden, as of Oct. 2007, about 10 percent of NSPs in the world have deployed MPLS-TE. Most of these NSPs are tier-1 NSPs, which can afford many talents in house to manage the MPLS system.

Diffserv-aware TE

Diffserv-aware TE means doing TE differently for different classes of traffic [RFC3564]. Conceptually, the physical network is divided into multiple logical networks, one per traffic class: High-priority class can only use its own logical network capacity while lower-priority classes can borrow capacity from the high-priority logical network if and only if the capacity is not used by high-priority traffic. Traffic Engineering is then done for each traffic class. The main purpose of Diffserv-aware TE is to avoid the concentration of high priority at any link. Otherwise, Diffserv may not work if there is a congestion of high-priority traffic.

Diffserv-aware TE introduces a considerable amount of complexity. Instead of one RSVP bandwidth value per link, each link now has multiple RSVP bandwidth values, one per class type. LSPs carrying different classes of traffic will need to be put into different queues. More information must be configured by the network operators, carried by the signaling protocols, and processed by the path computation algorithms. The part governing the dynamic borrowing of capacity among different logical networks is generally difficult to understand. It will require both Diffserv and TE knowledge to troubleshoot issues. Because of the complexity, Diffserv-aware TE has little deployment.

CDN

Simply put, Content Delivery Network (CDN) is to cache frequently accessed content in various geographical locations, and redirect access requests of such content to the closest place [Akamai]. Normally, CDN is considered completely separate from QoS. However, by moving content closer to end users, CDN can dramatically reduce delay, delay variation, and packet loss ratio for users' applications and thus their perception of network QoS.

Integration Challenge

Now that we have briefly described other traffic control schemes such as routing, TE, and CDN in addition to traffic management, we are ready to discuss the integration challenge. As we can see, whereas Diffserv traffic management is good for creating some service differentiation when there is congestion, routing and TE can be used to reduce congestion or prevent congestion from happening. CDN can be used to move content closer to the end users, thus making the delay, delay variation, and packet loss ratio of a long network segment that is left behind by CDN irrelevant. From this perspective, routing, TE, and CDN can be considered more effective for providing QoS than traffic management. Traffic management only affects the performance of traffic at a specific link of a network device. Therefore, it can be called a micro-control mechanism. Routing, TE, and CDN can affect traffic performance networkwide. Therefore they can be called macro-control mechanisms.

Routing, TE, and CDN complement traffic management well and should be used together. However, except for a few books and articles [Wang2][Xiao3], most QoS publications don't cover the coordination among these traffic control schemes. This is the integration challenge of the traditional traffic management-centric approach for delivering QoS. The consequence is that too much emphasis may be placed on a particular traffic control mechanism. For example, without considering TE for relieving congestion, one may feel that there is a strong need for traffic management. Consequently, too much traffic management mechanisms may be introduced into the network. Similarly, without considering CDN, one may feel that there is a strong need for TE and traffic management in the backbone to deal with a large amount of video traffic. Consequently, too much TE and traffic management mechanisms may be introduced, resulting in too much complexity in the network.

COMPLEXITY CHALLENGE

This section discusses the complexity challenge resulting from introducing various traffic control schemes to provide QoS.

Complexity/Control Spiral

When various traffic control schemes are introduced with the intention to provide better QoS to end users, additional complexity will also be introduced. On the one hand, the control schemes will allow the optimization of a network resource, for example, giving the limited link bandwidth to the high-priority traffic at the congestion point, or balancing traffic inside a network to avoid/reduce congestion. This will lead to better QoS. On the other hand, complexity will reduce reliability and lead to worse QoS. Therefore, there must be an explicit tradeoff between traffic control and reliability. In the traditional QoS wisdom, by separating QoS out from reliability, such a tradeoff is usually not considered. Reliability is taken for granted—somebody else will take care of it. This can lead to an over-favoring of various traffic control schemes, and can create a self-fulfilling QoS prophecy and a complexity/control spiral (originally called complexity/robustness spiral in [RFC3439]).

The self-fulfilling QoS prophecy (or to be exact, it really should be the self-fulfilling "traffic control" prophecy) refers to the fact that, in an attempt to sell new or additional equipment to service providers, network equipment vendors generally tout certain sophisticated traffic control features as differentiators. The vendors keep touting to the service providers about these mechanisms that the mechanisms actually stick in the service providers' mind. When the service providers issue Requests for Information (RFIs) or Requests for Proposal (RFPs), they will ask for such mechanisms just in case one day they will want to use them. Then the vendor folks feel vindicated: Service providers are asking for these mechanisms, they must be needed. A new cycle then begins, and both parties will become even more convinced.

If too many such mechanisms are deployed into the network, it will also create a complexity/control spiral, in which complexity creates further and more serious fragility, which then requires additional control knobs to deal with failure. The additional controls further increase complexity, and hence the spiral.

An example of such a complexity/control spiral is MPLS Fast Reroute [RFC4090]. Fast Reroute promises 50 ms traffic restoration like protection in SONET/SDH networks. If complexity is not considered, then Fast Reroute is obviously highly desirable. But one needs to understand how much complexity Fast Reroute introduces in order to provide such fast traffic restoration. Figure 8-1 illustrates providing Fast Reroute for a MPLS LSP that goes through five nodes.

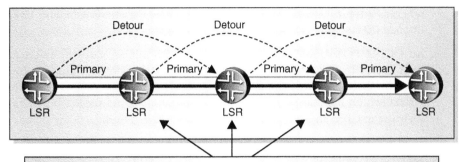

- Fast reroute is signaled to each LSR in the path
- Each LSR computes and sets up a detour path that avoids the next link and next LSR
- Each LSR along the path uses the same route constraints used by head-end LSR

FIGURE 8-1

Fast reroute illustration

One LSP becomes four LSPs (the primary and three detours). If the last link is to be protected, then there will be one more detour LSP from the second to last LSR to the last LSR via a different path. This increases the number of LSPs in the network and increases the signaling overhead. It is true that with "facility backup" [RFC4090], the detours can be shared among different LSPs, therefore, not every primary LSP will spawn three more detours. However, this introduces coupling among the primary LSPs through their common detour. Tight coupling among different components in a large system (i.e., the Internet here) generally makes the system more sensitive to small disturbances and permits more complications to develop, thus make the system hard to understand and more likely to fail [Doyle][Willinger]. The point here is that the pros and cons of Fast Reroute must be carefully weighed. Otherwise, Fast Reroute could potentially cause more fragility, and people may think that they need even stronger traffic restoration mechanisms.

Complexity Comparison between Internet and PSTN

Packet switching, as exemplified by the Internet, is considered by many people as simpler than circuit switching, as exemplified by the Public Switched Telephone Network (PSTN). After all, in a packet switching network, you don't need to set up a connection before sending packets. The success of the Internet may also lead people think that packet switching is better than circuit switching, and therefore must be simpler. But while it may be true that in the early days, the Internet (or ARPANET to be exact) was quite simple, today's Internet is no longer simpler than PSTN. This is because considerable complexity has been added over time, especially in recent years. The following observations should provide some perspective on this issue:

1. The typical software (that is, network operating system) for an Internet router requires between 8 to 10 million lines of code (including firmware),

whereas a typical packet switch requires on average about 3 million lines of code [RFC3439].

2. An OC192 POS router link contains at least 30 million gates in ASICs, at least one CPU, 300 Mbytes of packet buffers, 2 Mbytes of forwarding table, and 10 Mbytes of other state memory. On the other hand, a comparable circuit switch link has 7.5 million logic gates, no CPU, no packet buffer, no forwarding table, and an on-chip state memory [RFC3439].

3. During 1969 to 1999, about 2700 RFCs (in other words, technical specifications) were published by the IETF. In contrast, during 2000 to 2007, about 2300 RFCs were published. In other words, during the first 31 years, 54 percent of RFCs were published while in the last 8 years, 46 percent. Before and after year 2000, the fundamentals of the Internet did not change. Most of the newly added technologies are for optimization purposes. Some of them are fairly sophisticated and require considerable expertise to configure, e.g., color/admin group in TE, Diffserv-aware TE, FRR, etc.

The first two observations should serve to show that today's Internet is already more complex than PSTN, and the third should serve to show that complexity of the Internet can further surge. As we will see in the next section, complexity can have a big impact on reliability and QoS. Therefore, the tradeoff between the need for control for network optimization and the need to avoid complexity must be carefully considered.

Impact of Complexity on Network Reliability and QoS

The impact of complexity on QoS is twofold. First, increased complexity can lead to system fragility and thus more device failures. This causes higher packet loss ratio, delay, and delay variation. Second, increased complexity increases configuration complexity and leads to more configuration errors, which, in turn, leads to more network problems and lower QoS. According to a report from Gartner Group entitled "Making Smart Investments to Reduce Unplanned Downtime" [Scott], 80 percent of unscheduled network outages are caused by people or process errors. Therefore, we focus on the second and give a real-world example on how complexity can affect QoS. Our logic in this section goes like this: First, various QoS mechanisms can introduce complexity. We consider this self-evident; second, we use the example to show that even minor complexity can have a negative impact on QoS. These get us to the conclusion that QoS mechanisms can also have a negative impact on QoS (besides their positive effect).

Below is the example.

To make network management easier, NSPs usually assign a DNS name to each network interface (that is, one end of a link). The following are a few examples:

- chicr1-oc192-snvcr1.es.net
- so4-0-0-2488m.ar2.fra2.gblx.net
- csw11-fe4-5.sgw.equinix.com

- chi-bb1-pos1-0-0.telia.net
- ge-10-0.hsa3.Boston1.Level3.net

The names are usually long because they need to carry a lot of information. For example, chicr1-oc192-snvcr1.es.net will convey that this is an OC-192 interface from core router 1 (cr1) in Chicago (airport code CHI) to core router 1 (cr1) in Sunnyvale (airport code SNV). The link belongs to the U.S. Department of Energy's Energy Science Network (es.net). Similarly, so4-0-0-2488m.ar2.fra2.gblx.net is a SONET OC-48 interface at Global Crossing's access router 2 (ar2) at the second site in Frankfurt (FRA), and the port number of the interface is 4/0/0. Because the names are long, sometimes editorial errors happen, and the DNS names become incorrect. As an example here, let's assume that the network operator meant to enter ge-10-0.hsa3.Boston1.NSP1.net but mistakenly entered ge-1-0.hsa3.Boston1 .NSP1.net. In most cases, this had no consequence because the DNS name for the interface was just for informational purposes. So such mistakes might not be noticed. However, one day some customers called the network operations center to report packet loss. During troubleshooting, the network operator realized that there was a duplication of an IP address in the network. From the duplicated IP address, he did a reverse DNS lookup to trace it back to interface ge-1-0.hsa3.Boston1.NSP1.net, only to find out by surprise that the IP address was not a duplicate. Was a duplicated IP address the cause of the problem? There seemed to be no duplicated IP address at this interface. What's going on? By the time he realized that there was indeed an IP address duplication (at ge–10–0) and the wrong DNS name had pointed him to another interface (ge–1–0), 20 minutes had passed. Basically, the long interface name that consisted of 31 characters, a tiny complexity, led to a minor configuration error, which eventually led to an additional 20 minutes of packet loss for some customers. The packet loss time caused by the duplicated IP address was not counted here as it was a separate matter.

Now try the following configuration to complete a traffic management task of "map the PHB IDs to the appropriate traffic class/color combinations":

```
host1(config)#mpls diff-serv phb-id standard 0 traffic-class
best-effort color green
host1(config)#mpls diff-serv phb-id standard 10
traffic-class af1 color green
host1(config)#mpls diff-serv phb-id standard 12
traffic-class af1 color yellow
host1(config)#mpls diff-serv phb-id standard 14
traffic-class af1 color red
host1(config)#mpls diff-serv phb-id standard 18
traffic-class af2 color green
host1(config)#mpls diff-serv phb-id standard 20
traffic-class af2 color yellow
host1(config)#mpls diff-serv phb-id standard 22
```

```
traffic-class af2 color red
host1(config)#mpls diff-serv phb-id standard 46
traffic-class ef color green
```

For people that are not familiar with a router's Command Line Interface, the parts that really need to be entered by the network operator are just the numbers (that is, 0, 10, 12, etc.), the traffic-classes (af1, af2, etc.), and the colors (green, yellow, red). Other parts can largely be auto-completed. The real challenge here is to figure out the magic matching among the numbers, the traffic-class, and the color.

As if figuring out the matching is not daunting enough, the configuration manual further states:

> **Note:** This example includes both MPLS and policy configuration commands, and assumes that you are thoroughly familiar with the information and commands presented in the ERX Policy and QoS Configuration Guide.[1]

The "ERX Policy and QoS Configuration Guide" is only 106 pages long. This note would certainly make the network operator feel that the configuration task is a piece of cake.

If the magic matching is done wrong, for example, because of a typo, the network may exhibit some funny behavior. Now imagine another network operator has to troubleshoot the problem. Unless he is thoroughly familiar with the magic matching, he will have a hard time figuring out why.

For people who don't operate networks, "map the PHB IDs to the appropriate traffic class/color combinations" or "assign a DNS name to an IP interface" would sound very simple. These examples serve to show that even seemingly simple tasks can involve a fair bit of complexity, and even minor complexity can affect QoS. Therefore, the benefit of introducing various traffic control schemes—for example, Diffserv traffic management to create CoS—must be carefully weighed against the potential harm of additional complexity.

Finally, it is important to note that all NSPs understand that complexity is something to avoid. The real difficulty is to decide what is acceptable complexity and what is not. Currently, the quantitative relationship between Capital Expenditure (CAPEX), Operations Expenditure (OPEX), and a network's inherent complexity is not well understood. In fact, there are no agreed-upon and quantitative metrics for describing a network's complexity. So a precise relationship between CAPEX, OPEX, and complexity remains elusive. Complexity metrics for networks is therefore a good topic for further study. Before that happens, a rule of thumb may be that the network designers should talk with the network operations

---

[1] This example is not to pick on Juniper/ERX. Other vendors' CLI would involve a similar amount or even more complexity.

people before they decide to introduce a mechanism into the network. If the network operations people do not feel comfortable managing that mechanism, that is probably a sign of too much complexity.

INTEROPERABILITY CHALLENGE

There are two aspects in the interoperability challenge. One aspect is regarding the cooperation among NSPs so that they can sell end-to-end QoS. The other is regarding how a NSP can use equipment from different vendors to deliver QoS as planned.

Inter-Provider Interoperability Challenge

In Chapter 6, we discussed the lack of QoS settlement among NSPs and the commercial challenges associated with that. In this section, we will assume that such a settlement has been worked out somehow, and we will discuss some of the technical challenges regarding how NSPs cooperate.

The first challenge that NSPs may face is that they will likely be in different stages of supporting QoS. NSP2 may be willing to do its best to honor NSP1's QoS traffic. However, it may not have deployed Diffserv and won't able to differentiate traffic. This will reduce the end-to-end differentiation.

The second challenge is different NSPs may offer different numbers of traffic classes, and there is no standardized inter-mapping between different NSP's traffic classes. As we stated before, IETF only standardized the Per Hop Behaviors (PHBs) that can be used to construct different traffic classes. It didn't standardize the traffic classes themselves. Traffic classes are up to the NSPs themselves to define. Now let's assume that NSP1 offers COS's X, Y, Z and NSP2 offers COS's X, Z. If NSP2 maps NSP1's CoS Y its CoS X, the distinction between NSP1's COS's Y and X will become blurry. Similarly, if NSP2 maps NSP1's CoS Y its CoS Z, the distinction between NSP1's COS's Y and Z will become blurry. When the distinction between the two traffic classes become blur, the NSP will have difficulty selling the more expensive class.

The third challenge is how to evaluate whether a NSP that is involved in the end-to-end path delivers its promise on QoS. This may involve allocating the delay, delay variation, and Packet Loss Ratio (PLR) budgets. For example, in order for a VoIP call to work, the delay, delay variation, and PLR must be within a certain range, as described in Chapter 2. These are called the delay, delay variation, and PLR budgets. Now let's say that one VoIP call involves three NSPs (NSP1, NSP2, and NSP3), and the delay budget is 150 ms. Should the 150 ms budget be divided equally among the three NSPs, each getting 50 ms? This may not be fair if NSP1 covers a much bigger geographical area than NSP2 and NSP3. To make things even more complicated, let's say another VoIP call involves five NSPs. Instead of getting 50 ms, should NSP1 now get only 30 ms? Should the allocation be application session-specific? Or should the allocation be just NSP-dependent, that is, a NSP will get a certain delay, delay variation, and PLR budget no matter how many NSPs are involved in a specific application session?

Allocating a delay variation budget can be particularly challenging. While delay is additive (for example, a 100 ms delay in NSP1 and a 50 ms delay in NSP2 would result in a 150 ms overall delay), and PLR is approximately "additive" when it is sufficiently small (that is, below 1 percent), delay variation is not additive. A 10 ms delay variation in NSP1 and an 8 ms delay variation in NSP2 could result in an overall delay variation anywhere between 2 ms and 18 ms. Therefore, if an overall delay variation budget is 50 ms and three NSPs are involved, while it is acceptable to divide 50 ms into three parts, for example, 20 ms + 10 ms + 20 ms, and allocate them to the three NSPs, this can be an overkill, which will lead to over engineering of NSPs' networks.

[Y.1541] and [InterQoS] made some assumptions and discussed a number of options for delay/delay variation/PLR budget allocation. Some of these options may turn out to be feasible. However, none of them provides clean cut solutions. So this is a good topic for further research. In the end, this allocation challenge may not be as serious as it appears either, as we will see in the "Differentiation Challenge" section.

Regarding the inter-provider interoperability challenge, some people may argue that QoS (or more specifically, traffic management) need not necessarily be done end to end. One can also do it in a lightweight fashion, for example, just at the edge of the networks only or just at selective links. Therefore the inter-provider challenge may not be a big problem. This argument has some validity. However, although such a lightweight deployment may be practically useful, it may not be good enough to provide sufficient differentiation to sell CoS.

Inter-Vendor Interoperability Challenge

People that have been involved in the design of a router or switch know that the design and implementation of the QoS part (or to be exact, the traffic management part) is one of the most difficult parts. One major reason is lack of traffic management deployment in the field causes lack of feedback to the design team. As a result, if different design team members have different ideas, there is no authority to arbitrate. Consequently, implementations of the same traffic management function at different vendors, or at different products of the same vendor, can be slightly different. Sometimes, even the implementations at different parts (for example, a POS line card and an Ethernet line card) of the same system can be different. Below we give a few examples to show the challenge created by such difference. These examples are well known among the developer community but are not so well known among the user community.

Example 1. Different Handling of IPG in Shaping

Ethernet frames have clearly defined beginning and ending boundaries, or delimiters. These are marked by special characters and a 12-byte Inter-Packet Gap (IPG) that dictates the minimum amount of space or idle time between packets. Some shaping implementations take the IPG into consideration while some don't. As a result, to shape a user traffic stream to 10 Mbps, the shaper may be configured at 10 Mbps if the shaping implementation already excludes IPG, or more than 10 Mbps otherwise. In the second case, the exact value of the shaper has to depend on the average packet size, making it un-deterministic. This creates network management complexity. In the second case, even if the shaper is originally configured properly to some value over 10 Mbps, another network operator unaware of the IPG difference may think that it is a misconfiguration (because it doesn't match the user's traffic profile), and mistakenly change it back to 10 Mbps.

Because of such implementation differences, configuring a single device can already be challenging. Making devices from different vendors interoperable would be even more challenging.

Example 2. Different Handling of Bandwidth Sharing among WFQ Queues

A typically envisioned traffic contract involving multiple COS's, for example, a Premium class and a Best-Effort class, would generally go as follows. For the total physical pipe of 10 Mbps, Premium class can go up to 2 Mbps, and if there is no premium traffic, best effort can go up to 10 Mbps. This is normally enabled by a WFQ mechanism, in which two queues are configured, one with 20 percent link bandwidth and the other with 80 percent bandwidth. Traditional implementation of WFQ would allow the premium queue to get full link bandwidth if there is no best-effort traffic. After all, this is how WFQ is supposed to work. Some vendor's "advanced" WFQ implementation would disallow the premium question to get more than 20 percent of link bandwidth even if there is no packet in the best-effort queue. After all, this is what the SLA specifies. While one can argue which implementation is correct, in reality, both implementations exist. This again creates interoperability complexity.

Other challenges of configuring traffic management involve:

1. Lack of guideline on how to configure WRED parameters, for example, what discard probability should be configured at what average queue length. In some cases, this is not configurable. That is not necessarily a bad thing because few people know how to configure it anyway. But different vendors' settings can be different. Therefore, it is hard to make WRED really useful.

One would think that [RFC2309] entitled "Recommendation on Queue Management and Congestion Avoidance in the Internet" would give some guidelines on how to do such a configuration. But it didn't. It only provides some high-level guidelines as quoted below.

- RECOMMENDATION 1:
  Internet routers should implement some active queue management mechanism to manage queue lengths, reduce end-to-end
  latency, reduce packet dropping, and avoid lock-out phenomena
  within the Internet.

The default mechanism for managing queue lengths to meet these goals in FIFO queues is Random Early Detection (RED) [RED93]. Unless a developer has reasons to provide another equivalent mechanism, we recommend that RED be used.

- RECOMMENDATION 2:
  It is urgent to begin or continue research, engineering, and measurement efforts contributing to the design of mechanisms to deal with flows that are unresponsive to congestion notification or are responsive but more aggressive than TCP.

To a certain extent, this testifies to the lack of serious deployment of RED and WRED. Otherwise, there would be follow-up RFCs giving more specific instructions.

2. Lack of guidelines on how to configure the output rate of a queue, especially for a queue at a device in the middle of the network. See the "Technical Solution" section of Chapter 4 for more information on this topic.

3. Lack of statistics for various queues inside a network device, for example, average queue length, how many packets are dropped. Counters to provide statistics use mostly Static Random Access Memory, which is expensive. Consequently, there may not be sufficient counters provisioned during hardware design. This creates difficulty for the network operators to know what's going on, or how to fine tune their traffic management parameters. To draw an analogy, this is like trying to improve one's shooting skill in a completely dark room. After you shoot, you don't even know whether it's a hit or a miss, let alone how to improve.

ACCOUNTING CHALLENGE

Today, only the total bandwidth used at the user interface is accounted. If the user billing is usage based, then such a bandwidth will be used for billing purposes. There is no separate accounting of different classes of bandwidth. Separate accounting of different classes of bandwidth is doable, but as stated previously, it requires more counters, which may not be available in some network devices. Other tools related to accounting, e.g., statistics collectors, accounting tools, link utilization graphing tools, need to be upgraded to support multi-class accounting. All of these can be done but it requires a fair amount of work.

DIFFERENTIATION CHALLENGE

The most fundamental challenge of the traditional CoS-based approach to deliver QoS and sell QoS is the lack of attention to create a user-perceivable difference.

In many cases, people confuse creating differentiation with creating user-perceivable differentiation. While putting traffic into different queues with different priorities can certainly create some differentiation in delay, delay variation, and possibly packet loss ratio, such differentiation is not necessarily perceivable by the end users. If not, then such differentiation is not useful enough for selling CoS. Although using various traffic management mechanisms to enable the Differentiated Services PHBs may yield multiple traffic classes with perceivable quality differences in a controlled lab environment with artificial congestion, it may not be sufficient to create a user-perceivable difference in the real world. In this section, we discuss this important topic. For this discussion, the topics that we discussed in Chapter 2, i.e., what determines the end users' QoS perception, what the QoS requirements of user applications are, and what IP networks in normal condition can deliver, provide the necessary background.

Differentiation Difficulty under Normal Network Condition

In Chapter 2, we showed that under normal network condition, national networks as big as those covering the entire United States can meet the need of the highly interactive and delay-variation-sensitive applications. Intercontinental networks can meet the need of interactive applications, except in the less-developed regions. The typical PLR of IP networks in developed regions can also meet the need of most applications. This means that in the developed regions where the selling of QoS is intended, CoS may not be able to create much user-perceivable differentiation under normal network conditions—Best Effort itself is already good enough. This is somewhat intuitive because if there is no congestion in a network, then Diffserv CoS won't be that useful. While CoS may be useful in developing regions where capacity is in shortage, selling of QoS in those regions may not be a profitable undertaking.

Some people may argue that the current Internet in developed regions has good performance because the network utilization level is moderate, but network utilization will not always stay that way. They feel that they can cite many reasons to support their opinion:

- The coming of Internet video may eat up all the available link bandwidth and drive network utilization to exceed 70 percent, making CoS differentiation useful.

- NSPs want to have as high a network utilization as possible. It increases network efficiency after all. When economic downturn comes, cost will be tightly controlled, and network utilization may go up above 70 percent.

- One can never have too much bandwidth (that is, low link utilization). As long as idle capacity exists, either TCP will increase its sending rate to consume it, or new applications will be invented to consume it. The rapid increases in CPU power, memory capacity, and disk space over the years have all been consumed.

Table 8-1 Historic levels of average network utilization

Networks	Utilization
AT&T switched voice	33 percent
Internet backbones	15 percent
Private line networks	3–5 percent
LANs	1 percent

Because the network utilization level has a large effect on how to provide QoS and whether CoS is useful, we provide a discussion regarding what future network utilization will be like.

First, because history can provide good insight for us to predict the future, let's review the historic utilization of the Internet. Dr. Andrew Odlyzko, formerly a researcher of AT&T and currently Director of Digital Technology Center of University of Minnesota and Interim Director of Minnesota Supercomputing Institute, reported that historically most IP networks are lightly utilized (Table 8-1) [Odlyzko2].

The statistics here dated before 1998. At that time, web traffic had grown significantly but had not reached its peak. One may think that at the peak of the Internet bubble around 2000, the utilization could be higher. That's not the case. The Internet bubble actually created a glut of capacity. As a result, the link utilization was low and the price was low. It is true that after the bubble burst, no capacity was added for a period of time and demand caught up. However, from communications with network operators, link utilization did not appear to go up significantly. This is partly verified by a 2002 Merrill Lynch report that stated that for 2002, the average utilization of optical networks (all services) was about 11 percent, whereas the historical average before that was approximately 15 percent [ML2002]. While the exact network utilization level varies at different times in different networks, the key point is still very clear: Network utilization was fairly low before, during, and after the Internet bubble.

What may be more significant in the historic statistics is the extremely low network utilization for private line networks and LANs. Users have complete control of the desired utilization level they want. The fact that they decided to maintain such light utilization says a great deal about what future utilization will be like.

Now that we've discussed the historic network utilization level, let's move on to discuss future network utilization.

There are reasons to believe that network utilization will not go much higher than its historic level. The major factors leading to overprovisioning in data networks include:

1. Data traffic has been growing rapidly, no matter whether before, during, or after the Internet bubble. Because of that, and the difficulty of predicting

the exact growth, operators tend to add bandwidth proactively and aggressively.

2. Data traffic is perceived to be bursty. Link capacity is usually provisioned to handle the busiest hour on the busiest day. As a result, overprovisioning is the norm.

3. Falling price for coarser bandwidth granularity makes it economical to add capacity in large increments, especially with the wide adoption of Ethernet in the metro and wide area networks.

4. Internet traffic is highly asymmetric, but links are symmetric. Consequently, when capacity is added to address traffic growth in the busy direction, a large amount of overprovisioning results in the other direction.

With the historic utilization levels and the factors leading to overprovisioning in mind, we can now review the arguments that network utilization may go way up from today's level.

First, will Internet video significantly increase the link utilization of the Internet?

It's not easy to draw a conclusion on this. What we know is the rapid increase of web traffic starting from 1995 originally drove network utilization up. That attracted the attention of the world both commercially and technologically. Consequently, a large amount of capital was infused and technology was significantly improved to meet the demand. Consequently, the network utilization returned to its historic level. This has been confirmed by statistics reported in [Odlyzko2] and [ML2002].

But will the effect of Internet video be very different from the web's?

One possible way to answer this question is to compare the traffic increase factor of the web in 1996 and of Internet video today. Table 8-2 describes the typical bandwidth requirements of various applications [Caswell].

Table 8-2 Bandwidth requirement of various applications

Application	Speed requirement
Text	300 bps
Web browsing	Varies
Digital music	128–700 Kbps
Video conferencing	384–2000 Kbps
MPEG-4 VoD (Internet)	250–750 Kbps
MPEG-2 (DVD, Satellite)	4000–6000 Kbps
HDTV (1080i compressed)	20,000 Kbps

The speed requirement for web browsing is missing. For the following calculation it is assumed to be 10 Kbps. Note that the bigger the value is, the stronger our argument will be. Therefore we assume a relatively small value (only 20 percent of a dialup link), to hold ourselves to a higher standard.

So in 1996, compared to the previous popular application email, web browsing introduced a traffic increase factor of (10 Kbps/300 bps) = 33. Today, Internet video will introduce a traffic increase factor of (750 Kbps/10 Kbps) = 75. While the latter is about twice as high, the order of magnitude is the same. In addition, not every Internet video requires 750 Kbps. For example, the most popular Internet video portal (YouTube's video) requires 300 Kbps [Kleeman]. If we further consider that the number of users increased much faster around 1996 than today, the difference in total traffic increase happened in 1996 and happened today will be further reduced. Therefore, we can say that Internet video's traffic impact today will be comparable to the web's traffic impact in 1996. If the infrastructure could keep up then to maintain relatively stable link utilization, it is possible that it can also keep up this time. Note that some people may say that TV quality video requires more than 750 Kbps. That's true, but most of those traffics are broadcast in nature and can be served by a single copy of traffic (for all the users).

Second, is higher network utilization necessarily better? Two veteran network operators, Randy Bush and David Meyer, offered an opinion in [RFC3439]:

> *"As noted in [MC2001] and elsewhere, much of the complexity we observe in today's Internet is directed at increasing bandwidth utilization. As a result, the desire of network engineers to keep network utilization below 50 percent has been termed 'over-provisioning'. However, this use of the term over-provisioning is a misnomer. Rather, in modern Internet backbones the unused capacity is actually protection capacity. In particular, one might view this as '1:1 protection at the IP layer'. Viewed in this way, we see that an IP network provisioned to run at 50 percent utilization is no more over-provisioned than the typical SONET network. However, the important advantages that accrue to an IP network provisioned in this way include close to speed of light delay and close to zero packet loss [Fraleigh]. These benefits can be seen as a 'side-effect' of 1:1 protection provisioning.*
>
> *There are also other, system-theoretic reasons for providing 1:1-like protection provisioning. Most notable among these reasons is that packet-switched networks with in-band control loops can become unstable and can experience oscillations and synchronization when congested. Complex and non-linear dynamic interaction of traffic means that congestion in one part of the network will spread to other parts of the network. When routing protocol packets are lost due to congestion or route-processor overload, it causes inconsistent routing state, and this may result in traffic loops, black holes, and lost connectivity. Thus, while statistical multiplexing can in theory yield higher network utilization, in practice, to maintain consistent performance and a reasonably stable network, the dynamics of the Internet backbones favor 1:1 provisioning and its side effects to keep the network stable and delay low."*

In other words, link utilization higher than 50 percent is not necessarily better. Although it may lead to lower CAPEX, it may also lead to higher OPEX. This may be the reason why most NSPs maintain a moderate level of network utilization, and haven't put in the effort to drive it up.

Third, can people never have too much bandwidth (which is equivalent to moderate link utilization) because TCP is "greedy," or because new applications will always be invented to consume them all?

We will first discuss the effect of TCP. Some people think that it is not possible to properly overprovision a network, because whenever you add additional capacity, TCP's flow control mechanism [RFC2581] will lead applications to increase the sending rate to consume that additional capacity. This argument is not valid in a medium time scale, loosely defined as longer than one day but shorter than six months. The Internet is operated by many NSPs that add capacity at different times. A NSP adding capacity at a link to reduce its utilization does not remove all the communication bottlenecks. Because TCP's performance and sending rate are more constrained by those bottlenecks, the increase in TCP sending rate is generally much smaller than the link bandwidth increase. The result is that the NSP adding capacity will see its link utilization go down for at least a period of time.

We will then discuss the effect of new applications. Can people never have too much bandwidth, just like they never have too much CPU/memory/disk space? While it is true that people tend to need more CPU/memory/disk space over time, it is important to distinguish whether supply leads demand or follows demand. If supply leads demand, then utilization will be low most of the time. Otherwise, utilization will be high most of the time. In the case of CPU/disk space, it is clear that supply leads demand—most people's CPU and disk are at fairly low utilization. In the case of physical memory, it's the opposite and utilization is usually high. In the case of bandwidth, supply of backbone bandwidth tends to lead demand. Even in the last mile, supply leads demand more often than the other way around. Some evidence is (1) when web browsing could work with dialup modem at 56 Kbps, many people chose to sign up for broadband connections for faster speed. The typical utilization of last mile broadband links is fairly low (<10 percent) [Odlyzko2]; (2) NSPs often offer multiple tiers of broadband access speed, and many people don't sign up for the highest tier—their link utilization is already at a fairly moderate level at the non-highest speed. So while it may be true that people will want more bandwidth in general, their link utilization will stay at a reasonable level most of the time.

After we address the specific concerns that future link utilization may go way up, we offer a generic view why it will not from an economic perspective. In terms of total number of equipment and spending, the public Internet is relatively small compared to the total of all corporate networks [Odlyzko]. But corporate networks all have low network utilization, around 1 percent. This means that users are willing to pay to keep the network lightly loaded so that they can get good performance all the time (that is, even when they have a sudden burst of traffic). Therefore, if people were to vote on the desired network utilization for the Internet, they will likely vote with their wallet to keep the utilization moderate.

After all, they have voted to keep a much bigger network, that is, the total of all corporate networks, very lightly utilized.

In summary, it is likely that future IP networks will continue to have utilization below 50 percent (or at least below 70 percent). Many people familiar with IP networks would simply take this for granted. Therefore, the delay, delay variation, and PLR statistics presented in Chapter 2 will continue to hold true in the foreseeable future. In other words, in developed regions, Best Effort will continue to be sufficient for most applications in normal network condition. This means that it will continue to be difficult to differentiate COS's in the foreseeable future.

One point that is worth mentioning is some NSPs may offer one service over a PSTN network while another service over a packet network. For example, Verizon's current Ethernet Private Lines are provisioned over its next generation SONET network, while the Ethernet Virtual Private Lines are provisioned over a packet switched network [VZeth]. We agree that these two COS's can be differentiated to a point that the end users can perceive the difference. After all, PSTN circuits are more reliable and provide zero delay variation and packet loss. But we want to point out that this doesn't contradict our point here. We are saying that it would be difficult to differentiate two COS's in an IP network, not one in an IP network and the other in a PSTN network.

Differentiation Difficulty under Abnormal Network Conditions

Another value proposition of the traditional CoS-based approach for QoS is that when some link or node failure causes the network to be in an abnormal condition, CoS can be used to provide QoS for the high priority at the congestion point. When traffic management is the sole mechanism for enabling QoS, this value proposition is somewhat valid. However, if we take routing and TE into consideration, then the situation will be different.

First, if the failure happens at the last-mile links or the DSLAM uplinks: Because these links are the only connection between the end users and the network, the end users are going to lose connectivity. CoS won't be able to make a difference.

Second, if the failure happens outside the last-mile links and the DSLAM uplinks (note that DSL uplinks are the common bottleneck links): Routing or TE should be able to avoid congestion. This is because IP networks in developed regions typically have utilization well below 70 percent, and in most cases, well below 50 percent. This means that the network has more than 30 percent of its capacity reserved to handle failures or a sudden traffic surge. As discussed in [Odlyzko2], there is good reason for this and it is unlikely to change in the foreseeable future. Consequently, except for catastrophic failures that can cause the network to lose more than 30 percent or more of its capacity, routing or TE will be able to re-distribute traffic in a way that congestion can be avoided. In that case, CoS still won't be able to create user-perceivable differentiation. Note that the case where a network loses more than 30 percent of its capacity does happen.

But virtually all those cases are caused by software defects that cause widespread network device outage. In those cases, CoS won't be useful either.

Some people may argue that during abnormal network condition, after various traffic control schemes take effect to re-distribute traffic in the network, link utilization will go up. How do we know that Best-Effort service will still provide good enough performance at that time? Data traffic is known to be bursty after all. At high utilization, say 85 percent, will the burst create congestion, thus long delay, delay variation, and packet loss, in the sub-second time scale? If this is the case, then CoS can be useful to ensure high quality for priority traffic.

While NSPs monitor link utilization, the link bandwidth values are based on 1-minute or 5-minute aggregate. Therefore, while such link utilization graphs don't show 100 percent utilization, they cannot prove that micro congestion doesn't exist. To answer this question, it is important to note that there are well-conducted joint research works by Stanford University and Sprint Nextel Corp. showing that even at 85 percent link utilization, IP networks will incur only minimal delay, delay variation, and PLR [Fraleigh]. In fact, most of the network devices coming to the market after the Internet bubble burst will deliver minimal delay, delay variation, and packet loss ratio even at line rate (that is, 100 percent utilization) or close to line rate. There are plenty of third-party benchmarking results to prove that. Therefore, even if the network is running at 70 percent utilization, as long as the failure won't cause it to lose 15 percent of its capacity, it is fairly safe to conclude that Best Effort will continue to perform well. For a large network, 15 percent of its capacity is a lot of capacity. One or two failures are unlikely to cause the network to lose that much capacity. Granted that there could be cases in which a software defect would cause a widespread network outage. But in that case, CoS won't be able to create any differentiation either.

With all this said, it is acknowledged that there will be cases in which congestion does happen in a network and CoS can be very useful. The argument here is that such cases will be temporary and rare. Our fundamental assumption for saying that is, for competitive or other reasons, NSPs won't tolerate chronicle congestion anywhere in their networks. We believe this assumption is valid for at least NSPs in the developed regions. Given that, can NSPs sell CoS for service differentiation in those rare cases? We cannot say that it is impossible. But users who care that much for those rare disaster cases must have some critical applications that cannot tolerate any failure. Therefore, they will likely demand hard assurance, that is, good service quality at all times or a large refund. With a very limited number of customers and large risk, it is not clear that this business proposition will be commercially viable.

Poor Performance Happens but CoS Won't Be the Solution

Up to this point, we explained that it is difficult to differentiate COS's in both normal and abnormal network conditions. But some people may still doubt this. They may feel that while the Internet has pretty good performance in general, there are occasions when service quality is poor. So why can't those situations be improved with a higher CoS?

We agree that poor performance does happen occasionally, even in developed regions. But the question is, will CoS make a big enough difference in those scenarios to justify charging for it? To answer that question, we need to examine what causes poor performance.

The cause of poor performance could be at the end-point applications. For example, an overloaded web server could cause users' web sessions to be slow. CoS won't help there.

The cause could be at a congested last-mile link, either for a residential user or for an enterprise. We then need to ask why that happens, because as we discussed previously, last-mile link utilization is generally low. Also, in many cases, the NSPs will give the user an option for higher speed. So it is possible that the user decides to tolerate the bad performance to save money. CoS can help the user technically. But if the NSP wants to charge money, it won't work because saving money is the primary concern of the user at the first place.

The cause could also be at a congested non-last-mile link, for example, DSLAM uplink or peering link. Here the NSP could choose to upgrade link capacity or introduce CoS. CoS can possibly help on this link because application packets can be treated differently based on their sensitivity to delay, delay variation, and PLR. This may create perceivable differentiation to end users but there is no guarantee, because lack of differentiation elsewhere may blur the differentiation at this link. However, even if some perceivable differentiation is created, it would be very tricky for the NSP to charge for CoS. If the NSP does so, it becomes a perfect example of "evidence of poor quality." Users can charge the NSP for deliberately degrading network performance, by not upgrading link capacity, to create justification for charging for QoS.

The cause of poor performance could also be failure. When a network device fails, all the packets at that device will be lost. During the re-convergence of network control protocols, certain packets may be dropped too. Many network operators and equipment vendors know that a lot of today's network quality issues are caused by failures themselves instead of by the congestion after the failures. A joint study by AT&T Labs, Microsoft Research, and Yale University in 2006 concluded that IP networks have "Decent yet insufficient reliability," that "the (Internet2's Abilene) network is still far from achieving 99.999 percent reliability" [IPrelia]. A more realistic estimate on IP networks' availability is 99.9 percent. This amounts to about 9 hours of problems per year. But even the 99.9 percent availability may be an over estimate. Influenced by the notion that networks can have 99.999 percent availability, many network operators do not want to talk about their network availability in public if it is far lower than 99.999 percent. But we have seen internal documents from multiple Tier-1 NSPs on their network availability. One proudly stated that its optical transport network has 99.95 percent availability. Note that that is 50 times lower than the 99.999 percent availability. Others stated that their metro access networks have availability ranging between 98.5 percent and 99.5 percent (IP networks are generally less reliable than optical transport networks because they are more complex). Because the Internet consists of many IP networks, its reliability will be lower than individual IP networks. CoS is not going to help in reducing failures. One could even argue that it can increase failures because it introduces additional complexity into the network.

At the end of this section, after we explained the difficulty for CoS to create differentiation that is perceivable by the end users, it is important to say that CoS can be useful to create certain local differentiation. Our main argument is that such cases are temporary and rare, and the differentiation may not necessarily be perceivable to the end users. Therefore they are probably not sufficient to justify charging money for CoS.

SUMMARY

The key points of this chapter are:

- Traditional CoS-based approach for providing QoS focuses solely on traffic management, and omits the integration with other traffic control schemes such as routing, TE, and CDN. Traffic management only affects the performance of traffic at a specific link of a network device. It is a micro-control mechanism. Routing, TE, and CDN can affect performance of traffic network-wide. They are macro-control mechanisms.

- By treating QoS and reliability as separate issues, the traditional QoS approach can introduce too much complexity into the network because the complexity impact on reliability is not explicitly considered. This can lead to a complexity/control spiral where complexity makes the network more fragile and triggers the need for new control to handle that, which in turn introduces further complexity, and hence the spiral.

- Lack of interprovider QoS settlement can reduce incentive among NSPs to cooperate. Even if such a settlement exists, different NSPs may be at different stages of providing QoS; they can provide different numbers of COS's. The coordination itself can be tedious and time-consuming. This will make it difficult to differentiate COS's to a level that is perceivable by the end users. The allocation of end-to-end delay, delay variation, and packet loss ratio budgets among NSPs is also challenging. The implementation difference from different vendors for the same traffic management functionality makes the deployment of traffic management tedious and error-prone.

- The most fundamental challenge of the traditional CoS-based approach comes from the difficulty to provide user-perceivable differentiation. For developed countries, under normal network conditions, performance of Best-Effort service is already good enough for most applications. CoS won't create further perceivable differentiation. Under abnormal network conditions, routing and/or TE will likely be able to re-distribute traffic in the network to avoid congestion. CoS won't be able to create perceivable differentiation either. Furthermore, many quality issues during abnormal network conditions are related to the failure themselves, not the congestion afterwards. This further reduces COS's effectiveness in creating differentiation.

Up to this point, we have examined the commercial, regulatory, and technical challenges of the traditional QoS model. In the next chapter, we will summarize these key points and discuss the lessons.

Please voice your opinion about this chapter at http://groups.google.com/group/qos-challenges.

The Lessons

9

In this chapter, we will summarize the key points discussed in the previous chapters. By putting them close to each other and reviewing them in proximity, it is easier to see the theme. Also, these key points are derived with the generic QoS business model presented in Chapter 4. In this chapter, we will examine two specific QoS business models: QoS on Demand and Bandwidth on Demand. The purpose is to see what technical, commercial, and regulatory challenges discussed with the generic business model are still applicable in these specific models. Because these specific business models represent certain customizations to the generic model, this examination will help us see how certain customizations can affect certain challenges. This whole process will therefore provide some lessons on how to improve the traditional QoS business model to make it viable.

The most important points discussed in the previous chapters are:

- QoS is more than CoS and traffic management. CoS plus traffic management is just one possible way to realize QoS;

- It is technically challenging to differentiate multiple COS's. In normal network conditions, Best Effort already provides fairly good service quality especially in the developed regions such as North America, West Europe, and East Asia. There is little left to differentiate. In abnormal network condition, routing and TE can generally avoid congestion. Consequently, CoS can't provide much differentiation either;

- Lack of differentiation causes commercial challenges to sell CoS, because its value for the end users is not clear. If NSPs push to sell CoS, then resistance will appear and regulation issues such as Net Neutrality can arise.

These points are closely related. In fact, the first one leads to the second one, which in turn leads to the third one. This is the theme of this book.

QOS IS MORE THAN COS AND TRAFFIC MANAGEMENT

By defining QoS from an end-user perspective, that is, defining QoS as the capability to provide satisfactory experience for end users, QoS naturally becomes more than CoS and traffic management. Any schemes that can affect the end user's quality of experience naturally fall into the scope of QoS. This definition of QoS is productive because the ultimate goal of the network is to serve the user applications. This definition allows all important factors such as network capacity planning, traffic control, complexity, and reliability to be explicitly considered and traded off in the realization of the ultimate goal.

Some people may think the change of scope is no big deal. But the notion that QoS is more than Diffserv and traffic management actually has a far-reaching effect. In the past, when QoS was narrowly scoped within CoS and traffic management, people naturally focus on things inside this scope, and make assumptions for things outside this scope. Therefore, congestion is assumed to exist somewhere. Whether TE or other traffic control schemes can be used to relieve the congestion is not explicitly considered. As a result, Diffserv and traffic management would appear capable of creating some sort of differentiation. Whether this differentiation will be perceivable by the end users is not verified either. In contrast, when we broaden the scope of QoS and look at the whole picture, things become different. We will find that the "congestion exists somewhere" assumption is not really valid in the developed countries. Given that, can CoS and traffic management still make the differentiation? This leads us to the next key point.

IT IS DIFFICULT TO DIFFERENTIATE MULTIPLE COS's

The reason it is difficult to differentiate multiple COS's has been fully explained. It won't be repeated here.

It is acknowledged that there will be cases where congestion does happen in a network and CoS can be useful. From the discussion in Chapter 8, we concluded that such cases will be temporary and rare. Users who care that much for performance in those rare cases are likely to demand hard assurance, something that is difficult for NSPs to provide. Consequently, CoS may not be sufficient to make the business model commercially viable.

LACK OF DIFFERENTIATION CAUSES COMMERCIAL AND REGULATORY CHALLENGES

In the traditional QoS business model where QoS is sold as a separate entity from basic network connectivity (that is, QoS takes the form of CoS), the most fundamental issue is the difficulty of differentiating multiple COS's. It is this lack of perceivable differentiation that leads to the commercial and regulatory challenges.

First, because it's difficult to differentiate, the value of higher CoS is not clear. Given that, if a NSP attempts to sell a higher CoS to a user, that person will feel that either he (or she) won't be able to get any real benefit, or something is wrong with that NSP's Best-Effort service. The person may therefore consider switching to another NSP that provides "normal" Best-Effort service. Therefore, lack of user-perceivable differentiation is the fundamental cause of the commercial challenges.

Some NSPs (such as the broadband access providers in the United States) have a monopoly or close to monopoly market power. These NSPs can decide to sell CoS without much concern that their customers will switch to other providers. But if such NSPs attempt to do so, they will face a new set of challenges. This leads us to the second point.

Second, lack of perceivable differentiation, combined with forceful push by NSPs to sell CoS, leads to regulatory challenges. The Net Neutrality controversy is a case in point. If the ICPs and the consumers feel that they can indeed benefit from NSPs' CoS offering, they might not have taken the issue to the Congress but negotiated with NSPs on the specific payment terms instead. Or if they did take the issue to the Congress, their case would not have been so strong and gained so much support.

After summarizing the lessons that we learned from the previous chapters, we will move on to examine two specific QoS business models, QoS on Demand and Bandwidth on Demand. As we stated at the beginning of this chapter, the technical, commercial, and regulatory challenges discussed in previous chapters are for the generic QoS business model presented in Chapter 4. Every specific QoS business model represents a customization of the generic business model. These customizations may enable it to avoid certain challenges of the generic business model, or possibly introduce new challenges. In this section, we examine how the customizations in QoS on Demand and Bandwidth on Demand affect the challenges. This will provide some lessons on how to do further customizations, if necessary, to make the QoS business model viable.

PUTTING THINGS TOGETHER: DISCUSSION ON QOS ON DEMAND AND BANDWIDTH ON DEMAND

First we will explain what QoS on Demand and Bandwidth on Demand are. One commonly envisioned way to sell QoS is to give the consumers a "Turbo Button" at the selling NSP's web site. A consumer can click the button to buy QoS for his/her application(s) on demand. If the user's application performs better as a result, the user will remain in the turbo status and pay for the QoS he/she gets. If there is no perceivable difference in service quality, the user can terminate the turbo status at any time to stop paying for QoS. There are two variants in this approach. In the first one, when the user clicks the turbo button, some or all of his/her packets are prioritized. This is called QoS on Demand. In the second variant, only the access bandwidth is increased. For example, the DSL link between the user's DSL modem and the NSP's DSL Access Multiplexer (DSLAM) may be able to support up to 6 Mbps downlink, but the user's subscription can be just 1 Mbps. As a result, the NSP rate limits the DSL link to 1 Mbps. When the user clicks the turbo button before watching a TV-quality MPEG-4 movie online, the NSP may dynamically change the rate limit parameter to increase the user's DSL link speed. This is called Bandwidth on Demand.

QoS on Demand

Technically speaking, QoS on Demand can also include an increase of bandwidth. But to differentiate from Bandwidth on Demand, we limit QoS on Demand to only packet prioritization. This enables us to see the different effect of these two factors, packet differentiation and bandwidth increase.

The "double-selling" challenge of the generic QoS business model is still applicable but becomes less difficult. This is because in this specific model, the NSPs do not actively sell CoS but are letting the consumers decide themselves whether to buy CoS.

The "evidence of poor quality" challenge is applicable. In fact, how difficult the challenge will be depends on how successful the NSP is in selling QoS on Demand. If the users only buy QoS infrequently, then they may not be concerned about the QoS fee they pay, and may not consider switching to other providers. However, if the users have to buy QoS on Demand frequently, then they may become unhappy about the NSP's basic service, and may decide to switch to other providers that provide "better" basic service. In this scenario, a competing NSP whose service is at comparable quality may appear better just because they are not selling QoS. This is a dilemma for the NSP offering QoS on Demand—the more successful it is, the more likely the customer will defect.

The "who should pay" question is answered—whoever asks for it pays for it. This could be an enterprise, or more likely a consumer. What's not so clear is "will there be sufficient people willing to pay for it to make this model viable." In the end, this is reduced down to the capability to provide perceivable differentiation.

The "who should get" question is also answered—whoever willing to pay for QoS gets QoS. As we discussed before, this is not as good as "whoever needs QoS gets QoS." But arguably, only those people who truly need QoS for their applications will click the turbo button. Therefore, the glaring unfairness that rich people get QoS for their unimportant applications while poor people don't get QoS for their important applications no longer exists. Therefore, the "who should get" challenge is less severe.

As an on-demand scheme, the challenge associated with soft assurance, that is, users will not buy soft-assured QoS because they don't know whether it will make a difference, is no longer a problem. Because the users can decide at any time to stop buying, they can afford to try it out.

The "lack of inter-provider QoS settlement" challenge is still applicable. Lack of inter-provider QoS agreement will likely cause traffic prioritization to be enforced only locally. This will make the differentiation less perceivable, and reduce end users' incentive to buy.

On the regulatory side, if the service is offered solely to consumers, the Net Neutrality controversy may be less severe, as Net Neutrality is more concerned with NSPs charging ICPs for QoS than with NSPs charging consumers for QoS. If the service is also offered to enterprises, then the Net Neutrality controversy will remain.

On the technical side, the differentiation challenge remains because CoS remains the major means to realize QoS.

The customizations made in QoS on Demand help it avoid some challenges of the generic model, or reduce the severity. However, QoS on Demand also introduces some additional challenges. First, it should be noted that for this QoS charging model to be acceptable to the end users, the QoS charge has to be fairly low, for example, one dollar or less for an entire movie of about two hours. The reason is, viewing the movie itself will only cost the user three to five dollars, and the QoS fee has to be a fraction of that.

In summary, although QoS on Demand is able to avoid or alleviate some of the challenges, it cannot avoid or alleviate the most fundamental and difficult challenges, that is, difficulty to differentiate, possible perception as "evidence of poor quality," and the need for complicated interprovider cooperation. Its challenges are fundamentally similar to the generic QoS business model's.

Bandwidth on Demand

If a user's last-mile link's bandwidth is lower than the required bandwidth of an application, for example, a TV-quality MPEG-4 stream, the user simply cannot watch that movie. Bandwidth on Demand can possibly enable the user to do so by increasing the last-mile link speed. This is how Bandwidth on Demand can be useful. In contrast, differentiating traffic without increasing link speed will not be able to do so. In fact, as we explained in Chapter 2, increasing bandwidth on a low speed link has multiple advantages over introducing CoS. First, CoS just won't enable a user to watch a 2 Mbps stream in real time over a 1 Mbps link, no matter how you prioritize the traffic. In contrast, increasing the link bandwidth would solve the problem neatly. Second, with CoS, a high-priority packet still has to wait for a low priority that is already in transmission to complete. At a low-speed link, this waiting time is not ignorable. For example, waiting for a 1500-byte packet to finish transmission on a 1 Mbps link will take 12 milliseconds. This translates into an additional delay and delay variation of 12 ms. Its impact on 150 ms delay and 50 ms delay variation budgets is fairly big. In contrast, increasing the link speed to 3 Mbps would cut the wait to just 4 ms.

Next we will discuss what challenges of the generic QoS business model are applicable to Bandwidth on Demand.

The "evidence of poor quality" is no longer relevant, because there is no higher CoS to make Best Effort look bad. Plus, because how much bandwidth a user gets is fairly measurable, asking the user to upgrade from 1 Mbps to 3 Mbps would not make him feel that he is not getting 1 Mbps before the upgrade. In contrast, asking the user to upgrade from Best Effort to "premium" can make him feel that the NSP's Best Effort is worse than normal Best Effort.

The most fundamental challenge, lack of differentiation among COS's, is no longer relevant, because there is only one class.

The lack of interprovider QoS settlement challenge is no longer relevant. Existing peering/transit agreements among NSPs is based on bandwidth. That is sufficient.

The regulatory challenge is no longer relevant because there is no CoS and therefore no "traffic discrimination."

Other challenges for Bandwidth on Demand are identical to those for QoS on Demand.

In summary, although Bandwidth on Demand and QoS on Demand may appear similar, by detaching from CoS and focusing on bandwidth, Bandwidth on Demand avoids the fundamental and difficult challenges that QoS on Demand cannot avoid. To a certain extent, the conclusion would appear to be intuitive here: If it is difficult to differentiate COS's now and in the foreseeable future, then avoid CoS, or at least avoid touting it to the end users. This is the most important lesson in this chapter.

But Bandwidth on Demand is not without its own challenge. By enabling the users to purchase bandwidth only when they really need it, Bandwidth on Demand will likely prevent the NSPs from selling a higher-bandwidth tier as a subscription service to the end users. This point will be further explored in the next chapter.

SUMMARY

The key points of this chapter are:

- QoS is more than CoS and traffic management. By defining QoS from an end-user perspective as satisfactory network performance for end users' applications, any schemes that can affect the end users' quality of experience become QoS tools. This allows all important factors such as traffic control, complexity, and reliability to be explicitly considered and traded off.

- It is difficult to differentiate multiple COS's. Under normal network conditions, real-world measurement results showed that Best Effort can provide good enough performance for most applications in the developed countries. Under abnormal network conditions where failure causes a network to lose capacity, various traffic control schemes can re-distribute traffic to avoid or alleviate congestion. This can allow Best Effort to maintain good enough performance. Consequently, it is difficult to make a higher CoS perceivably better than Best Effort.

- Lack of differentiation among different CoS's leads to difficulty to sell QoS as a separate entity (that is, as a higher CoS). This is the fundamental cause of the commercial challenges. If NSPs push ahead to sell CoS anyway, users will defect to other NSPs. In the case in which the selling NSPs are a monopoly or duopoly, users will demand government intervention. That creates regulatory challenges for the NSPs.

■ QoS on Demand can avoid some of the challenges associated with the traditional QoS business model, but the most difficult and fundamental challenges remain. In contrast, by detaching from CoS and focusing on bandwidth, Bandwidth on Demand can avoid the most difficult and fundamental challenges.

From comparing the commercial viability of QoS on Demand and Bandwidth on Demand, it becomes clear that detaching from CoS makes Bandwidth on Demand more viable. To a large extent, this is an intuitive conclusion. If the most fundamental challenge is difficulty to differentiate, then we should try to avoid CoS, or at least not tout it. This gives us a direction in proposing improvement for the QoS business model, a topic that will be discussed in the next part of this book.

Please voice your opinion about this chapter at http://groups.google.com/group/qos-challenges.

The Next Step

This part contains five chapters:

- Chapter 10 proposes a revised pricing scheme for QoS, and discusses how it overcomes or relieves the most difficult commercial and regulatory challenges.
- Chapter 11 presents a revised technical solution and discusses its benefits.
- Chapter 12 presents two real-world QoS deployments and the lessons the operators learned.
- Chapter 13 discusses QoS in the wireless world.
- Chapter 14 concludes the book.

The New Business Model

10

In Part 2, from examining the challenges of the conventional QoS business model, we came to realize that many of its challenges are caused by the explicit selling of QoS on top of network connectivity. While CoS can be technically useful, the attempt to commercialize CoS creates the need to differentiate multiple COS's in a way that is perceivable to the end users. This is difficult to achieve. The lack of end-user perceivable differentiation in turn leads to other commercial and regulatory challenges. In this chapter, we propose an improvement to the conventional QoS business model. We then discuss how the new business model deals with the commercial, regulatory, and technical challenges. We also discuss why this model is better for the network industry.

An Internet QoS business model is essentially a pricing scheme for the Internet service. The proposed business model and the traditional business model each represent a specific pricing scheme. The effect of price schemes on the economics of the communication industry has been extensively studied and well understood by the economists. The lessons learned from those studies can tell us a lot about the possible evolution of Internet service pricing in general and QoS pricing in particular. However, many of the technical and management personnel in the Internet community are not aware of the conclusions from those studies. Therefore we will discuss in detail the historic evolution of communication service pricing schemes and their effect on NSPs' revenue. We hope that the conclusions and lessons there will make it clear why the proposed model is better suited for the industry.

The economic theory and many of the historic statistics presented in this chapter are reproduced from Dr. Andrew Odlyzko's research works with his permission [Odlyzko][Odlyzko3]. Dr. Odlyzko's contribution is acknowledged and his support to this book is greatly appreciated.

THE NEW BASELINE

Price QoS into the Services; Don't Sell QoS Explicitly

The gist of the proposed model is that NSPs should price QoS into the services they plan to sell, and just sell the services without touting QoS.

Note that a special case of this model is not to charge for QoS at all. In that case, NSPs will provide good service quality to all users at no extra charge. QoS essentially becomes a value-add or competitive advantage. It's up to each NSP on this.

In a nutshell, by pricing QoS into other services and not selling it explicitly, there will be neither commercialization of COS's nor the necessity to differentiate them. This removes the fundamental cause of the technical, commercial, and regulatory challenges. By not touting QoS, this also avoids setting up high user expectation and triggering scrutiny on service quality. As a result, the remaining challenges also become easier to handle.

We further propose that NSPs sell their services in the form of bandwidth blocks, with QoS priced in as appropriate. Each block will have a certain amount of bandwidth to enable certain applications. For example, the following three blocks can be used: (1.5 Mbps down, 384 Kbps up) for $19.99 per month, (3 Mbps down, 512 Kbps up) for $24.99, (15 Mbps down, 2 Mbps up) for $49.99. The first service can be called Internet access service. The second service will enable basic video service, allowing the user to watch MPEG-4 video, which generally requires 1.5 to 2.0 Mbps of bandwidth downstream. The third service will enable the user to get High-Definition TV (HDTV). This encourages users to pick the appropriate service (in other words, bandwidth block). VoIP may be supported in all three services. Similarly, block bandwidth pricing can be used for enterprises.

Commercial distinction between "basic bandwidth" and "premium bandwidth" is not proposed. Different services may have different amounts of bandwidth and possibly other differences, for example, different levels of technical support. But all bandwidth is commercially equal. This doesn't preclude NSPs from classifying and differentiating packets of different applications inside their network, for example, at a congested link when appropriate. But if such differentiation is done, it will be for the purpose of optimizing performance for the applications that are sensitive to delay, delay variation, and packet loss ratio. NSPs do not charge any money for doing it. Therefore, NSPs will not discriminate against any users because nobody pays extra. From an end-user perspective, what they care about is that the applications work in a predictable, consistent, and satisfactory way. They don't really care whether that is done via overprovisioning of bandwidth or via CoS. Therefore, this baseline serves that purpose well.

To draw an analogy, when buying a new car, buyers generally don't want to buy any extra option or extended warranty that the salespeople are eager to sell. The general perception is these are not essential and you will be ripped off if you buy them. Selling QoS on top of network connectivity in the traditional QoS business model is like selling the extra options or extended warranty. In contrast, if an option is included in the standard package, people are more willing to accept it. They don't even ask how much each option costs in the package.

From an economic perspective, the proposed model represents a pricing scheme that is based on bandwidth only. In contrast, the traditional model represents a pricing scheme that is based on both bandwidth and priority. Therefore, the proposed model represents a simpler pricing scheme than the traditional model. This is a very important point, and we will revisit it later in this chapter.

Currently, no residential users are paying for QoS. In this QoS business model, when a NSP is all ready to offer QoS in its network, it can price QoS into the contract for the next contract period. Today, in many countries, residential users usually have a yearly contract with their service provider. The contract monthly price can change after the initial promotion period, or at the contract renewal time. Users are already familiar with that. Therefore pricing QoS into the contract at the renewal time is feasible. The key is that there is no separate charge for QoS. If the original contract for a service is $19.99 per month, when the NSP decides to price QoS into the service, it can raise the monthly price to say $21. There is no option for the user to stick with the original contract and choose whether to pay $1 more per month for QoS. In fact, QoS is totally behind the scene.

Assume QoS is now enabled in the network, users that haven't got to the contract renewal time will also benefit from it—a little courtesy from the NSP. But they will also pay for QoS when they reach the renewal time. Regardless of QoS, users who sign up for the same service at different times can get different prices so this arrangement is not too different from existing business practices. The key of the proposed QoS business model is users need not be aware of QoS at all, and there is no discrimination of different users regarding QoS.

HOW THE PROPOSED MODEL DEALS WITH THE COMMERCIAL, REGULATORY, AND TECHNICAL CHALLENGES

Meeting the Commercial Challenges

In Chapter 6 we discussed the following commercial challenges for the conventional QoS business model:

- The "who should get" challenge
- The "double-selling" challenge
- The "evidence of poor quality" challenge
- The "what assurance to provide" challenge
- The "who should pay" challenge
- The "lack of interprovider settlement" challenge

Here we discuss how the proposed business model deals with these challenges.

The "who should get" challenge will be avoided. Because all people pay the same, there is no "unfairness" of "rich people get what they don't need while poor people don't get what they need." Some people may argue that with this model, it may become "nobody will get what they need" because the network will become congested. We argue that this is unlikely to happen. First, historic link utilization is fairly low, and there is good reason for that. Future link utilization is unlikely to be much higher either. These were discussed in detail in Chapter 8. One last point, although we propose no commercial differentiation, we don't preclude the usage of CoS for technical purposes, that is, handling of congestion, if needed.

The "double-selling" challenge will be avoided, because QoS is priced into the service. There is only one selling. That is, the selling of the services.

The "evidence of poor quality" challenge will be avoided. Since one is not selling QoS, its competitors cannot cite QoS as evidence of poor quality for the basic service. In fact, with QoS embedded, the service will have better quality.

The "what assurance to provide" challenge will be avoided. By not selling QoS explicitly, the proposed business model pretty much adopts today's assurance model, which is no assurance or soft assurance. In the traditional QoS business model, this is a problem because users may not buy QoS if there is no assurance or there is only soft assurance. But it is not a problem if we are not selling QoS explicitly.

The "who should pay" challenge will be avoided. Since the NSPs are not selling QoS explicitly, there is no issue of who should pay for QoS.

The "lack of interprovider QoS settlement" challenge will be avoided because it is largely related to CoS. Given that there is only one class of service, there is no additional interprovider challenge.

These should all be intuitive. But some people may argue that all commercial challenges go away only because we give up QoS. This concern will be addressed later in this chapter.

Meeting the Regulatory Challenges

Net Neutrality opposes traffic discrimination and particularly traffic discrimination against different application providers depending on whether they pay a QoS fee. Net Neutrality doesn't oppose NSPs from raising or lowering their service price as long as it applies uniformly to all businesses and people. Therefore, the proposed business model would not cause Net Neutrality controversy.

Meeting the Technical Challenges

There are two types of technical challenges for the conventional QoS business model. One type is related to the commercialization of QoS, for example, differentiating multiple COS's to make the effect of QoS perceivable to the end users so that QoS can be sold. The other type is related to delivering good service quality in an absolute scale.

With the proposed business model, the first type of challenge largely disappears. Since we are not selling QoS explicitly, there is no need to differentiate. As we discussed previously, among the technical challenges, the differentiation challenge is the most difficult one. It's avoided in the proposed model.

The second type of challenges remains the same. In other words, how to provide good service quality in a network in an absolute scale is the same with either business model. However, the users' expectation can be different. By not selling QoS explicitly, the proposed model avoids setting up high user expectation, which generally leads to close scrutiny of service quality. Therefore, the overall effect of the proposed model is, the second type of challenges also becomes less challenging.

THE POSSIBLE CONCERNS

While the proposed QoS business model may appear intuitive to some people, it may appear fishy to others. In this section, we address some of the possible concerns.

Are We Giving up QoS?

The first concern that may come to the skeptical people's mind may be: Is this a sophisticated way (or a conspiracy) to give up QoS? For example, what if competitive pressure prevents a NSP from pricing QoS into the service or forces it to price QoS at a nominal price? In that case, are we not giving up QoS?

We argue that the proposed QoS model doesn't necessarily lead to giving up QoS. To find out who is correct, we believe that it is equivalent to finding out which of the following scenarios is most likely to become reality:

- Scenario 1: basic network connectivity fee = QoS-enabled network connectivity fee
- Scenario 2: (basic network connectivity fee + QoS fee) = QoS-enabled network connectivity fee
- Scenario 3: basic network connectivity fee ≤ QoS-enabled network connectivity fee ≤ (basic network connectivity fee + QoS fee)

Note that the (basic network connectivity fee + QoS fee) arrangement represents the traditional QoS business model, while the (QoS-enabled network connectivity fee) arrangement represents the proposed QoS business model.

The skeptical people's argument is basically saying that the proposed model will lead to Scenario 1. That is, after QoS is provided and included in the network connectivity, it has to be given away for free because of competitive pressure.

In the remainder of this section, we will first explain that even if Scenario 1 happens, the proposed model is no worse than the traditional model. We then explain that if Scenario 2 or 3 happens, then the proposed model is better.

First, we acknowledge that it is possible that competitive pressure or other factors may indeed force a NSP to use QoS as a competitive advantage or as an internal cost reduction mechanism (e.g., to reduce the frequency of capacity upgrade).

That is, Scenario 1 can happen and QoS is given away for free. But under such competitive pressure, it is unlikely that a NSP will be able to sell QoS on top of basic network connectivity either. In fact, this has already happened to the traditional QoS business model. Today, there is little commercial QoS success as per the traditional QoS model. Plus, AT&T has agreed not to differentiate traffic in their broadband residential network for two years. This meant that with the conventional model, there will be no QoS revenue for two years. Therefore, the proposed model is no worse than the traditional model.

Second, if Scenario 2 happens, we argue that it is easier to sell QoS with the proposed model. This is because with the proposed model, there is only one sales transaction and there is no need to differentiate multiple COS's. This is fairly intuitive. And if we want to justify it beyond using intuition, there is an economic theory. This theory will be discussed in detail in the next section. Here we just provide a short summary. In economic terms, the proposed model represents a simpler pricing scheme—it is based on bandwidth only, not both bandwidth and priority. It is not that a simpler pricing scheme is always better than a complex pricing scheme. It is case dependent. But in the case of Internet service, because it is sufficiently cheap and is used frequently by common people, a simpler pricing scheme is better.

Third, if Scenario 3 happens, we argue that the proposed model will likely generate more QoS revenue. The reason is:

When QoS is sold separately in the conventional QoS business model, it is sold to enable premium services such as VoIP and video. In other words, QoS revenue will be derivative revenue of voice and video. Therefore, QoS revenue will likely be a fraction of VoIP and video revenue. However, VoIP is often free or close to free. Communications history also shows that online content distribution service such as online video/movie distribution will not generate much revenue either [Odlyzko4]. The reason is, the content distributor cannot charge the users much but has to give some of that revenue away to the content owners. For example, today's most successful content distribution company may be [Netflix], which has distributed more than 1 billion movies by postal service or online. The monthly subscription fee starts at only $4.99 and the most common subscription fee is around $10 to $15. This is for as many movies as one can view. Therefore, as a fraction of such revenue, QoS revenue as per the conventional QoS business model is likely to be a tiny fraction of a NSP's total revenue. After deducting the cost of the technical complexity of differentiating multiple COS's and complicated cooperation among NSPs, etc., the net profit of QoS will be even smaller. In contrast, with the new business model, when the NSP prices QoS into the service and raises the subscription price from $20 to $21, its broadband revenue increases 5 percent. This QoS revenue will likely be larger than the conventional model's. The key here is, when a NSP raises the monthly service price from $20 to $21, because both are sufficiently low and broadband Internet service is an important part of people's daily life, the users will not resist much. However, if a NSP sells QoS separately and gives the users a choice of whether to buy it, chance is that people won't buy it.

Without Explicit Selling of QoS, will Network Services Become Commodity?

Some people consider QoS as a premium on the top of basic network connectivity, which is a commodity. When QoS is priced into network connectivity, they are concerned that everything becomes commodity.

For various reasons, many people in the network industry prefer that network service be considered as a high-tech service instead of a commodity service. If something will turn the telecom service into commodity or move the industry towards that direction, these people will consider it bad. However, at the time this book is being written, the commodity companies are having record profit, especially the oil companies. Therefore being a commodity company is not necessarily bad from a business perspective. Eighty years ago, telephone service was used only by the richest people. Today everybody can afford it. So telecom service has been going in the direction of commoditization for a long time already. But in this process, the telecom industry continues to flourish and grow as a bigger and bigger component of the Gross Domestic Product (GDP). So commoditization is not unnecessary bad for the telecom industry either.

Therefore, the commodity question is an emotional question. It is not substantial from a business perspective. The more substantial question is "what does simple pricing scheme do to NSPs' profitability?". We will discuss that while addressing the next concern.

Will this Model Turn NSPs into Dumb Pipers?

The short answer is no. In the proposed model, NSPs can provide VoIP, video, and other premium services. These services are what the end users care while QoS is not. Therefore, from the end users' perspective, the NSPs are not getting dumber. What NSPs don't do is sell QoS explicitly. That's all.

Below we further clarify that simple pricing, dumb piping, and low profitability have no correlation. Examining the debate on Net Neutrality reveals that many people equate the three. It is not clear why but maybe the logic goes like this: Simple pricing is less sophisticated, i.e., dumb, and dumbness leads to lower profitability. Although this logic may appear intuitive, intuitive logic can sometimes lead to a wrong conclusion, as it does in this case.

First, simple pricing does not equate to dumb piping. As a matter of fact, the traditional "dumb pipers," i.e., the gas, electricity, and water companies, use usage-based pricing, a pricing scheme that is generally placed in the "sophisticated pricing" bucket. If NSPs use usage-based pricing, they would be charging users differently based on how much data the users downloaded and uploaded. This is not the case. NSPs' current pricing scheme is already simpler than the dumb pipers', and they haven't been called "dumb pipers."

Second, dumb piping does not necessarily lead to low profitability. Without proper government regulation and supervision, a "dumb" utility company can charge an astronomical rate for its commodity and record historic profit. Oil is a good example.

Third and most importantly, simple pricing does not equate to low profitability. Most tier-1 NSPs such as AT&T and Verizon in the United States use flat-rate or block pricing today, and maintain a healthy net profit margin around 10 percent. In contrast, the major airlines in the United States use differentiated pricing for economy, business, and first classes, and the majority of them are not nearly as profitable. Interestingly, the more profitable airlines are those that use a simpler pricing scheme. For example, Southwest Airlines is one of the most profitable airlines, and it only offered the economic class. Another example that simple pricing does not necessarily lead to low profitability is, in the third quarter of 2006, the Exxon Mobil oil company recorded a profit of $9.8 billion in the United States in a single quarter. It is not because the oil companies provided three grades of gasoline, but because the price of even the lowest grade (which most people use) was so high that profit went through the roof. In other words, even with simple pricing, companies can still achieve record profit.

Given that growth and profitability are most essential to the industry, and dumb piping and commoditization have no direct relevance to them, people need not be too emotional about the "dumb pipe" and "commodity" notions.

Are NSPs Forcing Their Customers to Subsidize the ICPs?

One important reason NSPs would like to charge ICPs a QoS fee is, the ICPs are using a lot of their network resource. Some ICPs such as Skype and Vonage even use the NSP's networks to provide a service that directly competes with the NSP's. Some of these ICPs have generated huge profit. Shouldn't they cough up a QoS fee to compensate for the huge amount of network resource that they consume? The conventional QoS business model would allow NSPs to do so—that is, if NSPs can somehow make the commercial, regulatory, and technical challenges go away. According to some NSPs, this would save the consumer from having to pay for QoS. But the proposed business model here would "force" all users, including the consumers, to pay. Isn't that a subsidy to the ICPs?

If with the traditional QoS business model NSPs were able to get the ICPs to pay for QoS so that the consumers didn't have to pay, then this argument would be valid. However, that didn't happen. Therefore, we can argue that the traditional QoS business model actually led to a lose-lose-lose situation for the NSPs, ICPs, and consumers—NSPs may not have incentive to provide QoS, while ICPs and consumers may not get good service quality. In contrast, with the proposed model, NSPs get QoS revenue while ICPs and consumers get good service quality. Because the ICPs use much more bandwidth than the consumers, they also pay more for the QoS that is priced into the bandwidth than the consumers. It is arguably a win-win-win situation.

Up to this point, we proposed a new QoS business model; explained that it can better handle the commercial, regulatory, and technical challenges; and addressed some possible concerns about the model. We believe that the arguments are intuitive. We also presented some simple evidences. But because the proposed model is a critical part of this book, we will further discuss its validity in detail.

The proposed QoS model and the traditional QoS model each represent a pricing scheme. Because we are not saying that the proposed model is the ultimate QoS business model, and we are only saying that the proposed model is likely better suited for the industry than the traditional model, we only need to show that the proposed pricing scheme is likely better than the traditional one.

Because neither model has become reality, discussing which one is likely better involves predicting the future. This is generally difficult. However, we would argue that one of the best ways to predict the future is to examine the historic trend. If there was a clear trend in history, and the current situation and the past situation are not fundamentally different, then there is a high likelihood that history will repeat itself in the future. This would give us some vision into the future.

In the next two sections, we will examine the evolution of pricing schemes in the 200-year history of the postal service and in the 100-year history of the telephone service. We will also examine how the pricing scheme affects the growth and profitability of these two industries. These two services are picked because they are both communications service, and are closer to today's Internet service than any other services. These examinations will reveal a clear trend. Therefore, if the trend favors the proposed model, then it would serve the purpose to show that the proposed model is more likely to prevail in the future.

THE HISTORIC PRICING TREND OF COMMUNICATIONS SERVICES

The historic pricing trend of communications services has been studied extensively. Numerous evidences showed that the trend is: Pricing scheme becomes simpler over time as the cost to provide such services decreases and the services become more widely used. This trend has been observed in postal service, telegraph, wireline telephony, wireless telephony, and Internet access.

In a sense, this is intuitive. When a new communication mechanism is invented, its price will be high and usage will be limited. The service will be mainly used by the government, by business, and by the upper class people. Although price is high, the cost is also high, and there is a lack of economy of scale. Consequently, the profitability is usually not high. As technology advances to bring cost down, price starts going down. The service attracts more people and the economy of scale develops. This further reduces cost and attracts even more users. A positive cycle is formed. Although the price of the service drops, usage usually grows so much that revenue becomes much higher, generally resulting in higher profitability. As the service price goes down, the pricing scheme generally becomes simpler because the price range for manipulation becomes smaller. This is not to say that service providers hadn't attempted to use complex pricing schemes to prevent the price from going down. Indeed, they had tried many times in history, and some will also try again and again in the future. This is just to say that over time, they will either willingly or unwillingly let price drop, adopt simpler pricing schemes, and see usage and profitability grow as a result.

Below we provide further analysis on this point. This analysis is largely a summary of a series of telecommunication pricing researches done by Dr. Andrew Odlyzko, Director of Digital Technology Center of University of Minnesota and a former researcher of AT&T. Interested readers can refer to his home page at http://www.dtc.umn.edu/~odlyzko/ for more information.

In [Odlyzko3], through parsing a large amount of statistics over a long period of time over a number of industries, a theory was formulated that pricing for a service is usually the result of two conflicting forces:

- The service provider's desire to maximize the value of its resource (for example, a telecommunication network), and
- The user's desire for value and simplicity.

If the former prevails, a sophisticated pricing scheme usually results. If the latter prevails, a simple pricing scheme results. The reason behind that is explained below.

For the same amount of resource, the service provider desires to give it to the users who value it most so as to maximize return. This generally calls for a sophisticated pricing scheme, such as priority-based pricing, distance-based pricing, time-of-day-based pricing, or some hybrid. There is an advantage to do this as it maximizes the efficiency of resource, albeit at the expense of additional complexity of metering by the service provider and extra record keeping by users to avoid unnecessary use of the resource. If the resource is expensive or hard to get, and the additional complexity is minimal and manageable, then a sophisticated pricing scheme that assigns the resources to the users who value them most will be beneficial to the whole society. Therefore it may prevail. However, if the resource is inexpensive, and is frequently used, then simplicity becomes a prominent factor in the dynamics. The users may even be willing to pay a little extra to obtain simplicity, because at a fraction of the base price, the extra will appear inconsequential to the users. A simple pricing scheme, e.g., flat rate pricing, will benefit the whole society and may therefore prevail.

For communications services, technology and competition are two important factors in determining which force will prevail. Technology is an important factor in determining the cost for providing communications service. For example, before programmed telephone switch was invented, telephone calls had to be manually switched by telephone operators. The cost for providing telephone service was much higher than today's. Similarly, the progress in semiconductor and WDM technologies significantly increases the capacity of the communications networks, thus reducing the cost per call. Competition, on the other hand, determines how much a user's desire will be respected. If a resource is monopolized and not regulated by the government, the service provider can have considerable power in deciding the pricing scheme, regardless of whether the resource is expensive or not. In contrast, excessive competition can lead service providers to sell their services below cost to court the users.

One may argue that government regulation is another important factor in determining which force prevails. There is validity in this argument. However, it should be noted that government regulation usually begins when the market cannot resolve the pricing issue in a mutually acceptable way by itself. From this perspective, government regulation can be considered as a second-order factor. A second-order factor can tip the balance only when the struggle between the first-order factors resulted in an impasse. However, because government regulation cannot be changed easily, once it is in place, it can change the outcome of the pricing struggle, as compared to a pure market-driven scenario. Therefore, government regulation can be indeed considered as the third determining factor.

Following this theory, the following conclusion can be drawn: If a service is expensive and used infrequently, then the desire for efficient usage of the resource will outweigh the desire for simplicity, resulting in a sophisticated pricing scheme. This is because the expensiveness of the resource justifies optimizing its efficiency while infrequent use makes the additional complexity for the users tolerable. However, if a service is inexpensive and used frequently, then the desire for simplicity will prevail, resulting in a simple pricing scheme.

There are examples in many industries to support this conclusion. For example, air transportation is expensive and is used relatively infrequently. The service is expensive because a service will cost at least several days of work[1] for common people. The service is used infrequently because on average people only travel by air two or three times a year [Odlyzko1]. Therefore, the airline industry ends up adopting a complex scheme to determine the price of a seat. The price can depend on class, distance, and time of purchase. Although this complex pricing scheme causes customer frustration, it is tolerated thus far. In comparison, bus transportation in a city is relatively inexpensive to provide, and the service is used frequently. A simpler pricing scheme, mostly flat rate or block pricing (with different block fees covering a different number of city districts), therefore prevails.

Below we show this theory at work in the postal and telephone industries. Both industries provide communication service. They are probably the closest to the Internet service.

Postal Service Pricing Became Simpler over Time

In the early days, postal service was expensive and was used infrequently. For example, in the 1830s, about fiver letters were mailed per person per year. In most parts of the country, the recipient had to learn that a letter was waiting, go to the post office, and pay a fee that cost about two hours of work on average [Odlyzko]. That would be equivalent to about $20 per letter today, as opposed to $0.42 in reality. Therefore, it was an expensive service then. As a result, pricing scheme was sophisticated in those days. The service was largely used by the government

---

[1] The reason we are using time-of-work instead of absolute dollar value to measure price is to avoid the need for inflation adjustment for a different era.

and by the upper class. In contrast, in 1994, there were about two pieces of mail per person per day [Odlyzko], or about 700 pieces per year, in the United States. The cost of each piece is now under two minutes' wage for a person at federal minimum wage. Table 10-1 depicts the pricing trend of the U.S. Postal Service rates for first-class mail.

From the table, it is clear that pricing schemes of postal service become simpler over time.

First, the distance dependency is reduced, from highly distance sensitive to moderately distance sensitive to practically distance insensitive. Today, even though letters destined to the same city supposedly cost less, most people simply use the regular stamp. It is sufficiently cheap that people don't want to bother finding out the exact in-city postage. This is an example that people are willing to pay a little more for simplicity.

Next, weight dependency is also reduced, although this is less obvious from Table 10-1 below. In the early days, letters we supposed to contain only 1 page, or the price would be much higher. This is where the word "single letter" in the table came from. Over time, the weight dependency is much reduced. Today, for common people, most letters are effectively within the 1 ounce limit. Weighing has become a rare practice.

Table 10-1 U.S. Postal Service rates for first-class mail

Year	Distance/Weight	Price	Price in hours of work
1799	Single letter		
	<40 miles	$0.08	0.8
	41–90 miles	$0.10	1.0
	91–150 miles	$0.125	1.25
	151–300 miles	$0.17	1.7
	301–500 miles	$0.20	2.0
	>500 miles	$0.25	2.5
1845	Single letter		
	<300 miles	$0.05	0.3
	>300 miles	$0.10	0.6
1863	First half-ounce	$0.03	0.2
1885	First ounce	$0.02	0.1
1999	First ounce	$0.33	0.02
2006	First ounce	$0.39	0.02

Telephone Service Pricing Became Simpler over Time

In the early days of telephony, the service was very expensive and was used infrequently. For example, in 1900, basic monthly service in New York City cost about $20 per month, which is equivalent to $2000 per month today. The pricing scheme was complex and many factors affected the price. These factors included the distance between the two calling parties, the duration of the call, the time of the day when call was made, the arbitrage between countries, and in the early days even the distance between the subscriber's location and the telephone switching office. In certain countries, the telephone companies published books, which contained nothing but the price per minute between any two cities, and that was the only way for anybody to figure out the price. Over time, technology advancement brought the cost down, allowing more people to use the telephone. Consequently, pricing became simpler, too.

First, distance between the subscriber's location and the telephone switching office has become irrelevant. Although it cost that telephone companies more to provide a line to a farther location, the service installation fee was averaged over a large number of subscribers. Therefore the price dependency on subscriber location was removed.

Second, the price sensitivity to distance between the calling parties was reduced. For example, Table 10-2 shows that price sensitivity to distance became less and less over time. Starting from 1997, the charge of U.S. domestic long-distance calls became distance-insensitive.

Third, the sensitivity of time of day was reduced. In 1919, there were three different rates for telephone services: the highest from 4:30 a.m. to 8:30 p.m., a lower rate from 8:30 p.m. to midnight, and the lowest from midnight to 4:30 a.m. Today, time-of-day variation was reduced to at most two tiers, with many calling plans offering uniform rates around the clock.

Fourth, local calls in some countries, e.g., the United States, became duration-insensitive. A flat monthly fee is charged for local service instead.

Therefore, it is also clear that pricing of telephone service becomes simpler over time.

Table 10-2 Price of a 3-minute call from New York City

Year	Philadelphia	Chicago	San Francisco
1917	$0.75	$5.00	$18.50
1926	0.6	3.4	11.3
1936	0.5	2.5	7.5
1946	0.45	1.55	2.5
1959	0.5	1.45	2.25
1970	0.5	1.05	1.35
2005	0.3	0.3	0.3

Now that we have examined the historic pricing trend of the postal and telephone industries, and the theory behind the pricing evolution of communications services, we have established some basis for predicting the pricing trend for Internet service. Today, broadband Internet access costs around $25 per month. For most households, this represents less than 1 percent of the income. Therefore we can say that the service is inexpensive. At the same time, email and web access have become part of people's daily lives. In other words, the service is used very frequently. Therefore, by the theory, one can conclude that charging for QoS separately is unlikely to happen. It would represent a move towards a more complex pricing scheme than today's, that is, from priority-independent to priority-dependent. In contrast, pricing in the proposed model remains the same as today's pricing scheme. It is more likely to happen. From an intuitive perspective, given that broadband Internet service has become sufficiently cheap, users don't want the burden of having to decide from time to time whether QoS is worth paying for. They would most likely prefer a simple pricing scheme to give them the peace of mind, even if they have to pay a little more for each unit of bandwidth.

This may also explain why many individuals outside the telecom industry, including some artists, signed the petition to support Net Neutrality. They may not understand QoS or communications economics, but they may intuitively detest moving towards a more complex pricing scheme.

Now it should become clear that pricing will become simpler over time. But some telecom professionals may still feel that "If only we can charge for QoS, more money will flow into the telecommunication industry and it will be good for the industry." In the following section, we will examine the correlation between pricing and usage, and show that simple pricing schemes can stimulate growth. Therefore, simpler pricing schemes can benefit the industry more than complex pricing schemes.

THE CORRELATION BETWEEN PRICING SCHEME AND USAGE GROWTH

In the previous section, we showed that pricing schemes of the postal and telephone industries become simpler over time. In this section, we examine how usage and profitability evolves over time as the pricing schemes become simpler.

Evolution of Postal Service Usage

Table 10-3 shows that coming with continuous price reduction and a simpler pricing scheme is continuous usage and revenue growth for the postal industry. Usage went from 0.8 million pieces of mail in 1790 to almost 200 billion in 1998. Expenditure went from 0.03 million in 1790 to almost 57.8 billion in 1998. Although there have been many predictions that mail would decline as a result of competition from telegraph, then telephone, and most recently email, it continues to expand to date.

Table 10-3 Growth of the U.S. Postal Service

Year	Pieces of mail (millions)	Pieces of mail per person per year	Expenditures (millions)	Expenditures as percentage of GDP
1790	0.8	0.2	$0.03	0.02 percent
1800	3.9	0.73	0.214	0.05
1810	7.7	1.07	0.496	0.09
1820	14.9	1.55	1.161	0.18
1830	29.8	2.32	1.933	0.21
1840	79.9	4.68	4.718	0.28
1850	155.1	6.66	5.213	0.2
1860			14.87	0.39
1870			24	0.33
1880			36.54	0.35
1890	4005	63.7	66.26	0.51
1900	7130	93.8	107.7	0.58
1910	14,850	161	230	0.65
1920			454.3	0.5
1930	27,887	227	803.7	0.89
1940	27,749	211	807.6	0.81
1950	45,064	299	2223	0.78
1960	63,675	355	3874	0.77
1970	84,882	418	7876	0.81
1980	106,311	469	19,412	0.7
1990	166,301	669	40,490	0.7
1998	197,943	733	57,778	0.68

Profitability information is not available for the U.S. Postal Service. But one can imagine that as transportation costs decreased, technology for sorting mail improved, and the economic scale became larger, postal service became more profitable over time. Table 10-4 shows such evidence for the postal service in the United Kingdom.

One interesting point to note is a large portion of today's mail is "junk mail." These are new usages of the postal system that were not foreseen in the early days when postal service was expensive. The "junk mail" phenomenon conveys

Table 10-4 British Postal Service volume and financial statistics

Year	Letters (millions)	Revenue (millions)	Profit (millions)
1711	$0.56	$0.45	Data not available
1760	1.15	0.42	Data not available
1770	1.53	0.78	Data not available
1780	2.09	0.68	Data not available
1790	2.86	1.66	Data not available
1800	5.42	3.6	Data not available
1810	9.28	5.69	Data not available
1820	10.96	6.93	Data not available
1830	11.33	6.52	Data not available
1839	75.9	11.95	Data not available
1840	168.8	6.8	2.5
1841	195.5	7.48	2.81
1842	208.4	7.89	3
1843	220.5	8.18	3.2
1844	242.1	8.53	3.6
1845	271.4	9.48	3.81
1846	299.6	9.82	4.13
1847	322.1	10.91	4.92
1848	328.8	10.72	3.7
1849	337.4	10.83	4.2
1850	347.1	11.32	4.02
1851	360.6	12.11	5.59
1852	379.5	12.17	5.45
1853	410.8	12.87	5.87
1854	443.6	13.51	5.98
1855	456.2	13.58	5.33
1856	478.4	14.34	6

an important point: When price becomes sufficiently low and the pricing scheme becomes sufficiently simple, creative new usages will appear (although the junk mail creativity is annoying to many people). This will happen to Internet communication service as well and will bring new growth.

Another important statistic to note is the postal expenditure as a percentage of the Gross Domestic Product (GDP). It grew from 0.02 percent of the GDP in 1790 to almost 0.9 percent in 1930. Since then, it has declined to about 0.7 percent [Odlyzko3]. However, this number is for the postal service alone. It doesn't account for the revenue of the express service companies such as [UPS] and [FedEx] that appeared after 1930. If we add up the revenue of these companies, we will find out that the portion of GDP contributed by mail services is still growing to date. The lesson here is: As price drops and pricing schemes become simpler, usage increase allows revenue to grow continuously over a time span of more than two centuries. Growth in usage often triggers new types of usage, too, for example, junk mail and express parcel delivery. Combined with dramatic cost reduction from economy of scale and technology advance, the postal industry has been able to thrive for more than two centuries.

Evolution of Telephone Service Usage

Table 10-5 shows that along with continuous price reduction and a simpler pricing scheme came continuous usage and revenue growth for the telephone industry. Revenue went from $16 million in 1890 to $246 billion in 1998. Even more revealingly, the revenue as a percentage of the GDP grew from just 0.12 percent in 1890 to 2.88 percent in 1998.

Table 10-5 Growth of U.S. fixed telephone service

Year	Revenue (millions)	Revenue as percentage of GDP	Phone calls per day (millions)	Phone calls per person per day
1890	$16	0.12 percent	0.0015	0.00002
1900	46	0.26	7.8	0.1
1910	164	0.46	35.6	0.4
1920	529	0.58	51.8	0.5
1930	1186	1.32	83.5	0.7
1940	1286	1.29	98.8	0.7
1950	3611	1.27	171	1.1
1960	8718	1.73	288	1.6
1970	19,445	1.99	494	2.4
1980	59,336	2.13	853	3.8
1990	133,837	2.33	1272	5.1
1998	246,392	2.88	1698	6.3

Table 10-6 Time spent online by AOL users

Calendar quarter	Average time per customer per day (minutes)
2Q1996	13
3Q1996	14
4Q1996	**21**
1Q1997	**31**
2Q1997	**38**
3Q1997	40
4Q1997	41
1Q1998	46
2Q1998	44
3Q1998	47
4Q1998	48
1Q1999	55
2Q1999	52
3Q1999	55
4Q1999	57
1Q2000	64

The revenue numbers in Table 10-5 do not include cellular voice revenue, which is even bigger than fixed voice revenue today. Cellular voice is a new type of usage grown out of fixed voice. The above revenue numbers do not include Internet access revenue either, which is another new type of usage grown out of fixed telephony. If the revenue of cellular voice and Internet access is added up, then the growth of revenue and revenue as a percentage of the GDP for the telephone industry is even more significant.

Some people may wonder whether usage naturally grows over time or if it is indeed correlated with a pricing scheme. The following usage statistics for America On Line (AOL) should help to answer that question (Table 10-6).

In October 1996, AOL changed its pricing scheme from usage based to flat monthly rate. In that quarter and the subsequent quarter, usage jumped. Before and some time after the pricing change, the usage increased much more slowly. This showed that pricing has a significant effect on usage.

HOW COMPARABLE ARE POSTAL AND TELEPHONE SERVICES TO INTERNET SERVICE?

Some people may argue that postal and telephone services are not comparable to the Internet service. Taking telephone service as an example, the people who don't think that they are comparable to Internet service will usually cite three differences:

1. Telephone service billing is usage based. The more minutes you call, the more money the service provider will get. Internet service is not. It adopts a flat monthly rate.

2. There is a natural upper bound for telephone usage. One can't talk 24 hours a day but one can run a peer-to-peer (P2P) application downloading a movie at home while working in the office.

3. Telephone service consumes little bandwidth. Internet service can consume a lot of bandwidth if a P2P application is used.

While it is true that telephone service and Internet service may appear quite different, the real difference is much smaller. Regarding the three differences above:

1. While telephone service billing has been usage based for a long time, and many telephone service providers very much prefer it that way, it has pretty much become a flat rate service in many countries for local and long distance calls. The only exception is international calling. But the people who have a lot of international calls usually use a calling card plan that is very cheap. To give some perspective, for $1, people can call from the United States to China and talk for 1 hour at toll quality. This may sound incredible but there are plenty of such calling plans on the market these days. Therefore, the telephone service providers might actually make more money by raising the monthly rate somewhat and allow users to make international calls. In other words, from a billing perspective, telephone service pricing is getting more and more "flat rate". It is not as different from Internet service as some people may think.

2. It is true that people don't talk around the clock while a P2P application can download around the clock. But because we are discussing trend, we must look at things over a long period of time. While it is true that some people can use the Internet a lot more than they use the telephone, it is also true that people today use the telephone a lot more than they used it in the past. In the early days, only rich people used the telephone, for important and urgent things only, and for as short a time as possible. While today some NSPs may feel disgusted that some people use their Internet access to download content around the clock, the telephone service providers in 1920 would feel equally as disgusted if people used their telephone to call their friends to chat, a common phenomenon today. They would consider telephone chatting unacceptably wasteful. In 1934, Bell System fined

one of its customers for using his telephone to call the fire department to report a neighborhood fire [Odlyzko]. That's considered by the Bell System as a breach of contract because the customer was supposed to use his telephone for his own business only. So people use their Internet service more than they use their telephone. But they also use their telephone more today than they used to. From a trend perspective, it is not too different.

3. Telephone service generally consumes less bandwidth than Internet service, but network capacity in the Internet era is also much larger, and much cheaper, than network capacity in the telephone era. Again, if we take a historic perspective, for a long period of time, it cost the NSPs more to provision for a 64 Kbps voice call, than for the NSPs to provision for a 1 Mbps download session today, given that network capacity has gone up by far more than 16 times. Therefore, the difference in bandwidth consumption is not really that big a difference either.

Another point of comfort is, the difference between the postal service and the telephone service is arguably bigger than the difference between the telephone service and the Internet service. Therefore, given that the trends in the postal service and the telephone service are decidedly similar, we can say that the trend in the telephone service will probably be applicable to the Internet service. This serves as further evidence that the trend comparison among postal service, telephone service, and Internet service is relevant, as all of the services are for communication purposes.

WHY THE PROPOSED MODEL IS GOOD FOR THE INDUSTRY

In this section, we summarize the key points expressed in this chapter, and point out why the proposed QoS business model is likely better for the telecommunication industry.

Fewer Commercial, Regulatory, and Technical Challenges

By pricing QoS into the network service and not selling it explicitly, the proposed QoS business model can avoid the most fundamental and difficult commercial challenges of the conventional QoS business model. Although some other commercial challenges can't be avoided, those are the less severe ones.

By pricing QoS into the network service and not discriminating users' traffic based on whether they pay a QoS fee, the proposed model avoids the Net Neutrality controversy, and therefore the regulatory challenge.

By not selling QoS explicitly, the proposed model avoids the need to differentiate multiple COS's in a user-perceivable way, which is required in the conventional model in order to convince users to buy QoS. This avoids the most difficult technical challenge.

Better Stimulus for Usage Growth

The proposed QoS model represents a simpler pricing scheme than the traditional QoS model. Communications history indicated that when a service becomes sufficiently cheap and frequently used, its pricing generally becomes simpler, not more complex. Therefore the proposed model has a better chance of being adopted. Communications history also indicated that a simpler pricing scheme generally simulates usage growth and increases the service provider's revenue and profit. For example, simple pricing in the postal industry helped to stimulate mail advertisement, which adds significantly to the revenue of the postal industry. Similarly, the flat monthly rate of local phone service in the United States helped to stimulate the early adoption of the Internet service in the United States because users need not worry about an additional local service charge caused by their dialup Internet access. Therefore, if the historic trend continues to hold true, the proposed model can result in higher revenue and more profit than the conventional model.

Better Focus on Real Revenue Opportunities

CoS commercialization is a distraction for NSPs. It attracts a lot of attention, time, and effort without producing real benefits. Consequently, it also diverts attention away from investigating and investing in other revenue opportunities such as peer-to-peer and Web 2.0 applications. By clarifying the commercial viability of selling CoS, this book will hopefully help NSPs focus on other real revenue opportunities.

THE EARLY EVIDENCE

Today, only a small number of NSPs succeeded in generating revenue from providing QoS. A common theme among these NSPs is that they emphasize the service they provide, not QoS. In particular, most of these NSPs just price QoS into their service and don't sell QoS separately. Therefore, their QoS business model is very similar to the proposed model. Two examples are provided below.

Akamai is a company that provides CDN service. Its selling point is that it will help Internet Content Providers (ICPs) provide better user experience. Akamai does so by caching certain content in its own servers, which are distributed all over the world, therefore effectively moving such content closer to the end users. This generally leads to lower delay, delay variation, and PLR. Many online video/movie providers such as Google and MediaZone use Akamai's service. Akamai sells its bandwidth at a higher price than other NSPs, because its bandwidth is effectively QoS enabled, although they don't necessarily put QoS in the spotlight.

Internap, in its own words, "provides Internet service with QoS to enable other companies to migrate business-critical applications—such as e-commerce, streaming audio/video, Voice over Internet Protocol (VoIP), video conferencing, virtual private networks and supply chain management—to the Internet." Internap does so by connecting to multiple tier-1 NSPs, and picking the best-performing Internet path via one of these NSPs to deliver its customer's Internet traffic. Internap sells its bandwidth at a higher price than other NSPs'. QoS is priced into the Internet service and is not sold separately.

Some people may ask a question from a completely different angle. They may argue that today, most of the incumbent NSPs are already using block bandwidth pricing so "what is new in this proposal?" To answer that question, we would like to point out that this book is not concerned with whether we have an innovative idea. We are concerned with arguing that selling QoS explicitly and separately from the network services themselves is not productive, and block bandwidth pricing with QoS priced in is a better alternative. If people readily accept that, it is great. But we believe that there are some NSPs who are considering selling CoS/prioritization. Therefore, advocating our point has merit.

THE POSSIBLE CUSTOMIZATIONS

In the future, if the Net Neutrality controversy settles in a way that is completely favorable to the NSPs, it is possible that certain customizations can be made to the baseline QoS business model. This topic is for further study.

Independent of that, it is expected that some NSPs will attempt to roll out some complex pricing scheme, e.g., one that is based on priority. There were numerous attempts in communications history by NSPs to use complex pricing schemes to segment the market as much as possible. But note again that communications history has clearly shown that such attempts were generally unproductive. Pricing schemes will become simpler over time, especially when a service becomes sufficiently cheap and frequently used.

SUMMARY

The key points of this chapter are:

- The proposed business model for QoS is that service providers should price QoS into their services and not sell QoS explicitly.

- By pricing QoS into the network services and not selling it explicitly, the proposed QoS business model can avoid many of the commercial challenges of the conventional QoS business model.

- By pricing QoS into the network service and not discriminating users' traffic based on whether they pay a QoS fee, the proposed model avoids the Net Neutrality controversy, and therefore the regulatory challenge.

- By not selling QoS explicitly, the proposed model avoids the need to differentiate multiple COS's in a user-perceivable way, which is required in the conventional model in order to convince users to buy QoS.

- Not charging for QoS explicitly does not mean that NSPs give up QoS, or that NSPs will become dump pipers. There is no correlation among simple pricing, dumb piping, and low profitability. Communications history indicated that as a service becomes sufficiently cheap and frequently used, its pricing scheme tends to become simpler over time. The simpler pricing scheme will in turn lead to usage growth and higher profitability. This trend is clear for the postal and telephone services. If it continues to hold true, it is likely that the proposed QoS model will prevail over the traditional model.

- There is early evidence that the proposed QoS business model will be adopted. Akamai and Internap's practices are examples of that.

This chapter deals with commercial and regulatory aspects of QoS. How to provide QoS in a network will be the topic of the next chapter.

Please voice your opinion about this chapter at http://groups.google.com/group/qos-challenges.

The New Technical Approach

11

In Chapter 8, we described the following technical challenges of the traditional traffic-management-centric QoS approach:

- Integration challenge: lack of consideration on how to make use of various traffic control schemes such as routing, TE, etc. to complement traffic management; this causes an overemphasis on traffic management.

- Complexity challenge: lack of explicit tradeoff between the benefit of various traffic control schemes and their complexity; this causes a plethora of controls and complexity in the network, and reduces its reliability.

- Interoperability challenge: lack of commercial QoS settlement among NSPs; this causes coordination difficulty among NSPs to differentiate COS's end to end.

- Accounting challenge: lack of accounting support on a per CoS basis; this causes difficulty in commercializing CoS-based QoS.

- Differentiation challenge: lack of end-user perceivable differentiation between Best Effort and higher COS's in developed countries; this causes difficulty in selling CoS.

With the new QoS business model, CoS is transparent to the end users, if used at all. Therefore, the challenges above that are related to CoS, that is, accounting and differentiation challenges, become irrelevant. The interoperability challenge also becomes largely irrelevant because it is largely concerned with CoS differentiation. NSPs will still need to cooperate. But little coordination is needed. All they need to provide is the best possible service. This was insufficient with the traditional CoS-based approach, because every NSP trying to provide the best possible service does not necessarily create the needed differentiation among COS's.

In this chapter, we propose a technical approach for delivering QoS in the revised business model, and describe how it addresses the remaining challenges: the

171

integration and the complexity challenges. This technical approach consists of the following steps:

- Network planning, to make sure that the network is properly provisioned for normal condition and possible failure condition; the purpose is to avoid congestion as much as possible.

- Network auditing, to reduce human errors so that such errors won't cause performance issues for the end users.

- Traffic control, to avoid congestion as much as possible, or to handle congestion if it is not avoidable.

- Traffic optimization, to proactively change the traffic matrix in one's network to optimize performance for end-user applications.

- Performance measurement, to find out how one's network really performs and whether any improvement is needed. This step, together with the previous ones, forms a closed control loop for the QoS delivering process.

This technical approach highlights network planning, network auditing, and performance measurement on the top of various traffic-control schemes. The philosophy is to use capacity as the primary means to provide QoS, and use various traffic controls as the ancillary means. To draw an analogy, the philosophy is to avoid fire as much as possible, but at the same time keep some fire extinguishers in the house just in case.

Before we discuss each of the steps in the subsequent sections, it should be pointed out that there can be no one-size-fits-all QoS solution. This chapter's intention is to present a generic solution framework that is relevant to most NSPs. Specific customizations will be needed for each NSP before the solution can be deployed.

NETWORK PLANNING

Network planning is to plan and provision one's network so that it meets current and future demand in a cost-effective way. Here cost-effectiveness is from a Total Cost of Ownership (TCO) perspective. That is, both CAPEX and OPEX are considered. For the same traffic demand, maintaining 50 percent average network utilization will imply higher CAPEX than maintaining 80 percent utilization. However, the operations can be simpler and it may lead to a lower OPEX. Therefore, what utilization level is most appropriate for a NSP should be carefully planned.

In the traditional traffic-management-centric approach, network planning is considered as something independent and is not explicitly discussed in providing QoS. This is not very practical because depending on the amount of capacity provisioned in network planning, traffic management may not have any significant effect on QoS (for example, in an overprovisioning scenario). Just like prevention is the most effective cure, good network planning is more effective for providing QoS than any specific traffic control schemes. Therefore network planning is discussed in detail here.

Generally speaking, network planning involves:

- Understanding current traffic demand.
- Predicting traffic growth trend.
- Dimensioning one's network to meet current and future demands.
- Planning for the failure scenarios.

The following sections address these aspects.

Derive Traffic Matrix, Analyze Traffic Trend

A traffic matrix is a two-dimensional matrix with its *ij-th* element t_{ij} denoting the amount of traffic sourcing from node i and exiting at node j.

	1	2	...	j	...	N
1						
2						
...						
i				t_{ij}		
...						
N						

Here nodes $1, 2, \ldots, N$ are the nodes in the network. Each node can be an individual router, or a site that contains multiple routers. Traffic matrix can represent peak traffic, or 95th percentile traffic, or traffic at a specific time. Depending on whether the network is to be provisioned based on peak traffic or 95th percentile traffic or something else, the appropriate traffic matrix can be used.

The traffic matrix clearly depicts how much traffic enters the network, where, its distribution inside the network, and at what places the traffic exits the network. It is the ultimate traffic map of one's network.

A traffic matrix is important for network planning. Meaningful network planning requires a traffic matrix.

While one may think that every NSP knows its traffic matrix, it is not the case. Traffic matrix is actually quite difficult to obtain. The reason is, for the network of N nodes, there are N^2 elements (in other words, unknown variables) to determine for the traffic matrix. Mathematically, this requires N^2 conditions. However, NSPs generally only know the amount of traffic entering each node from outside, traffic exiting from each node to outside, and traffic on the internal links. Roughly speaking, because few networks are fully meshed or close to fully meshed, there are

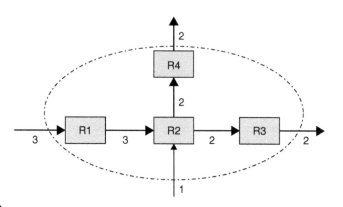

FIGURE 11-1

Traffic distribution in the network

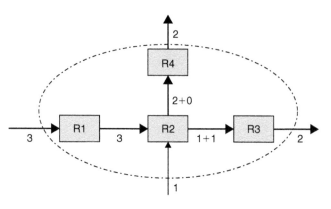

FIGURE 11-2

Traffic matrix scenario 1, all R2 traffic goes to R3

generally fewer than N^2 conditions. Consequently, it is not possible to determine all the N^2 elements of the traffic matrix analytically.

The following example illustrates the point above, and why a traffic matrix is needed for accurate network planning.

With the traffic map depicted in Figure 11-1, both scenarios in Figures 11-2 and 11-3 are possible. Note that traffic in the other direction is assumed to be 0 and ignored in the pictures.

In other words, the above example illustrates that the traffic matrix is undetermined – it is not clear how much R2 traffic goes to R3 and how much goes to R4. Without the knowledge of a traffic matrix, it would be difficult to do accurate network planning. For example, if it is known that traffic entering R2 from an external network is growing quickly, it won't be clear whether this is going to add pressure to the link to R3 or to the link to R4.

With additional information, it is possible to derive the traffic matrix. For example, if a full mesh to MPLS-TE tunnels is created among all routers, then the tunnel

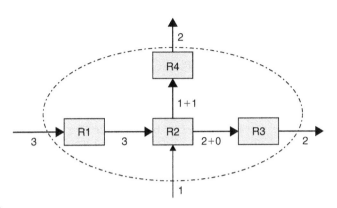

FIGURE 11-3

Traffic matrix scenario 2, all R2 traffic goes to R4

statistics will naturally provide information on the amount of traffic between each router pair, and thus the traffic matrix. Alternatively, [Netflow] could be used. At a high level, this involves:

- Enabling Netflow on all interfaces that source/sink traffic into the network.
- Exporting data to central statistics collector(s).
- Calculating the traffic matrix from source/destination information.

MPLS TE mesh and Netflow can provide accurate traffic matrix, but both involve a large amount of work. If these are not feasible, there are still other approaches to estimate the traffic matrix. Again, the more information is available (remember we need N^2 conditions), the more accurate the estimation result will be, but the more work will be involved. For the current best practice for obtaining the traffic matrix, one can refer to [TraMatr]. For subsequent discussions, we simply assume that the traffic matrix is obtained somehow.

To analyze traffic trends, historic traffic statistics and other information (for example, customers' new orders or traffic forecast) can be used. For example, if there are a series of traffic matrices for each month in the past twelve months, denoted by $A, B, \ldots, L$, then

- The trend matrix from A to B is $T_1 = B * A^{-1}$
- The trend matrix from B to C is $T_2 = C * B^{-1}$
- The trend matrix from K to L is $T_{11} = L * K^{-1}$

Using $T_1, T_2, \ldots, T_{11}$ and other information (new customer orders, forecast, etc.) that is available to the NSP, the NSP can estimate the trend matrix for the future months. For example, it could be a weighted average of $T_1, T_2, \ldots, T_{11}$. With the current/past traffic matrices and future trend matrices, the NSP can also estimate future traffic matrices. These traffic matrices can then be used for network planning.

Plan Network Capacity

Traditionally, capacity planning involves monitoring link utilization and adding capacity to the highly utilized links. Future traffic demand, if known, is translated into an additional load on the links in determining which links will have high utilization. This translation is largely done in an empirical way. Unless utilizations of different links differ significantly, routing policy changes are not attempted. If they are, they will be done in a trial and error basis on the live network, which could affect performance of network traffic and cause customer satisfaction issues. In general, without some network design tools and traffic matrix information, it is difficult to do capacity planning in a more scientific way than this.

With traffic matrix and network design tools from [WANDL], [Cariden], or [OPNET] a lot of simulations can be done offline. For example, these simulation results can tell network operators

- The simulated link utilization, which can be compared with network reality to check for accuracy.
- The simulated future link utilization, which can be used for capacity planning.

These tools also enable many what-if analyses and the determination of counter measures for certain scenarios. For example, what if traffic grows 50 percent in the next six months? What if a critical link fails? What if a link promised by a third-party carrier's carrier cannot be delivered in time? What if I don't add capacity but simply change the routing policy to achieve better traffic distribution in the network? Many such what-if analyses still involve estimation, for example, of the new traffic matrix. But as we will see in the next section, such estimation can be done in an iterative fashion to increase accuracy. Because all of these can be done offline, network operators can take the time to do a more thorough investigation before the final change is applied to the network. Consequently, capacity planning becomes more scientific.

Plan Routing Policies

Routing policies are the top-level criteria for making routing decisions. Both IGP and BGP have a decision process in picking the best route for forwarding a packet. Deciding routing policies basically means deciding IGP and BGP decision parameters to achieve the desired routing result.

The IGP decision process is relatively simple. It basically picks the path with the lowest IGP metrics, which is the sum of the metrics for all the links in the path. Therefore, deciding IGP policy basically involves deciding the IGP metrics for the links in the network. In general, NSPs like to divide their networks into a few regions and confine traffic originating from a region inside that region as much as possible. This divide and conquer approach creates some routing hierarchy and makes traffic control a little easier. This can be achieved by setting the

IGP metrics of the inter-region links one or two orders of magnitude higher than any intra-region link so that even the longest intra-region path is shorter than any inter-region link.

The BGP decision process is rather complex. It can be vendor specific too. Below is the BGP decision process of Cisco's routers [BGPBest]:

1. Prefer the path with the highest WEIGHT, where WEIGHT is a Cisco-specific parameter.
2. Prefer the path with the highest LOCAL_PREF.
3. Prefer the path that was locally originated via a network or aggregate BGP subcommand or through redistribution from an IGP.
4. Prefer the path with the shortest AS_PATH.
5. Prefer the path with the lowest origin type. Origin type denotes whether the path is originated from IGP or BGP.
6. Prefer the path with the lowest Multi-Exit Discriminator (MED).
7. Prefer eBGP over iBGP paths.
8. Prefer the path with the lowest IGP metric to the BGP next hop.
9. When both paths are external, prefer the path that was received first (the oldest one).
10. Prefer the route that comes from the BGP router with the lowest router ID.
11. If the originator or router ID is the same for multiple paths, prefer the path with the minimum cluster list length.
12. Prefer the path that comes from the lowest neighbor address.

This complex BGP decision process is beyond the comprehension of most people. Even inside a NSP, only a few people fully understand this process and have the privilege to manipulate such parameters. Suffice to say here that deciding BGP routing policies means deciding how to manipulate the important decision parameters such as WEIGHT, LOCAL_PREF, AS_PATH. This process is largely driven by commercial reasons. For example, for a small NSP that purchases transit services from two large NSPs, if one large NSP charges less for bandwidth than the other, the small NSP will prefer to send traffic to it as much as possible. This can be achieved by setting the BGP routing policy accordingly.

Designing optimal BGP routing policy can be more an art than an engineering process. Luckily, optimal solution is not necessary. Trial and error is usually involved in finding an acceptable solution. This process is done rather infrequently too, only at the beginning of a network, or after some significant changes. One can refer to [RoutArch] for more information on how to do this.

It should be noted that BGP and IGP routing policies are closely related to the traffic matrix. A traffic matrix depicts traffic distribution in a network under given routing policies. When BGP routing policy changes, the exit points of certain traffic will change, therefore the traffic matrix will change too. Even IGP routing policy change can possibly lead to change in the preferences for different peering points to the same next-hop AS, and can therefore cause traffic matrix change.

On the other hand, routing policies can be influenced by the traffic matrix as well. While initial routing policies are usually determined on a trial-and-error basis, partly because the traffic matrix is not available at the beginning, the determination of routing policies and the traffic matrix can be done in an iterative fashion. That is, first, given whatever user traffic information, one decides the IGP and BGP routing policies based on one's knowledge and experience, and sees how well the traffic is distributed in the network. Then the routing policies can be tuned until acceptable traffic distribution is achieved. After that, the traffic matrix can be obtained. With the traffic matrix, one can use a simulation tool such as [Cariden] or [WANDL] to fine-tune the routing policies until the most desirable distribution is achieved. Theoretically, when the routing policies change, the current traffic matrix may no longer be applicable. Therefore, the theoretical correctness of this approach is unproven. How to do this in a theoretically correct way can be a good topic for further research. But intuitively, because the routing policies are only fine tuned, the real traffic matrix under the fine tuned routing policies would likely be very similar to the current one. Therefore, this iterative process should work. In reality, it is usually considered working because the iterative process need not find the optimal solution: As long as it results in some improvement, it is considered working.

Plan for the Most Catastrophic Events

With network design tools, network operators can simulate all kinds of single-point failures and double-point failures to see how well the network can perform under such failures. If the network will perform badly under certain failure, counter measures can be planned or taken ahead of time to prevent such scenarios from happening. For example, if some part of the network will become badly congested under some failure, capacity may be added ahead of time. Alternatively, a new routing policy can be determined for such a failure scenario by simulation so that it can be immediately applied when such a failure happens. This will prevent congestion from happening or greatly reduce the congestion duration. In summary, by planning carefully ahead of time, networks can become more robust. Consequently, fewer complex traffic control schemes may be needed to deal with failures. This forms a positive cycle where robustness and simplicity enforce each other, as opposed to the complexity/control spiral described in Chapter 8.

It should be noted that the traditional CoS-based approach benefits less from such careful planning, because the planning will generally result in all COS's performing better under the failure scenarios, and thus insufficient differentiation. This is so unless the planning is done to make sure that under the failures higher CoS will perform well but lower CoS won't. Such "deliberate negligence" of Best Effort is not only more complicated to do, but also more prone to customer outcry and possibly lawsuits.

Security Consideration

Generally speaking, network security factors can play an important role in determining the QoS of a network. A network that succumbs to security attacks will not be able to deliver satisfactory performance. Security issues can be at the data plane, as well as the control plane. At the data plane, security should ideally be done end to end starting from the end users. Therefore, NSPs should encourage their end users to encrypt the data themselves, if encryption is desired. Encryption at the NSPs while leaving the user data in clear format at the access links only has limited value. At the control plane, protocol hand shake and authentication should be deployed as much as practical. External audit is also encouraged [Arbor].

A survey of current practices for securing service provider networks can be found in [OPsec]. A survey of standards efforts related to network security can be found in [SecEfforts]. A set of best practices for cyber security and physical security can be found at http://www.nric.org/. It is advisable that network operators stay informed and follow the best practice.

NETWORK AUDITING

Like network planning, network auditing is also important for providing QoS. Many service-affecting events are caused by operator errors. In Chapter 8, we described an example where a seemingly harmless DNS error created hurdles for the troubleshooting of a duplicated IP address error. This caused the service-affecting event to last longer. In fact, according to a survey done by David A. Patterson [Patterson], renowned professor of computer science at University of California at Berkeley, operator error is by far the largest cause of system failures, which leads network QoS issues (Figure 11-4).

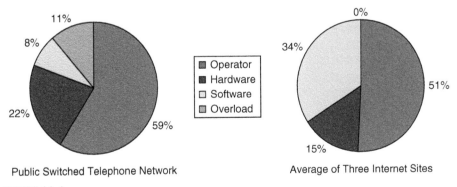

FIGURE 11-4

Percentage of failures by operator, hardware, software, and overload for PSTN and 3 Internet sites

Therefore, by finding out and correcting human errors as much as possible, networks will have fewer issues and thus better QoS.

Because virtually all NSPs export their router configuration files and other statistics (for example, link utilization) every day to a server, auditing can be done offline based on those configuration files. This makes auditing feasible, even if it is computationally intensive and time-consuming.

In addition, many service-affecting events have warning signs before they become catastrophic. For example, before a large-scale security attack is fully launched, some abnormal things will usually happen. For example, port scanning activity may increase dramatically, and the utilization of certain user links may go up precipitously. By detecting the warning signs and taking preventive actions, some catastrophes can be prevented or their detrimental effect can be much reduced.

Check for Misconfigurations

Networking devices are sophisticated devices and their configuration can be quite complicated. A typical configuration file of a NSP router has thousands of command lines or more. While many phrases can be auto-completed during the configuration, the key parameters must be entered by the network operators. To get the parameters right requires not only knowledge about the subjects (e.g., Diffserv or WRED) but also the specific representation of the parameters in the Command Line Interface (CLI). The following examples illustrate the challenges. The first example is to "map the PHB IDs to the appropriate traffic class/color combinations":

```
host1(config)#mpls diff-serv phb-id standard 0 traffic-class
best-effort color green
host1(config)#mpls diff-serv phb-id standard 10 traffic-class
af1 color green
host1(config)#mpls diff-serv phb-id standard 12 traffic-class
af1 color yellow
host1(config)#mpls diff-serv phb-id standard 14 traffic-class
af1 color red
host1(config)#mpls diff-serv phb-id standard 18 traffic-class
af2 color green
host1(config)#mpls diff-serv phb-id standard 20 traffic-class
af2 color yellow
host1(config)#mpls diff-serv phb-id standard 22 traffic-class
af2 color red
host1(config)#mpls diff-serv phb-id standard 46 traffic-class
ef color green
```

As one can see, it is not trivial to figure out why 10 is associated with traffic class af1 with color green, and 20 is associated with traffic class af2 with color yellow.

The next example is to enable WRED on the interface and specifies parameters for the different IP precedences:

```
interface Hssi0/0/0
   description 45Mbps to R1
   ip address 10.200.14.250 255.255.255.252
   random-detect
   random-detect precedence 0 32 256 100
   random-detect precedence 1 64 256 100
   random-detect precedence 2 96 256 100
   random-detect precedence 3 120 256 100
   random-detect precedence 4 140 256 100
   random-detect precedence 5 170 256 100
   random-detect precedence 6 290 256 100
   random-detect precedence 7 210 256 100
   random-detect precedence rsvp 230 256 100
```

Again, it is not trivial to make sense out of these parameters. When there is a problem, it is also hard to figure out whether the problem is caused by misconfiguration, or by implementation defects, or by something else.

Given the complexity in configuring the networking devices, it can be good to have a peer review of network configurations before they are applied to the devices. This not only helps to identify potential errors but also helps to reduce troubleshooting time, should another network operator have to do the troubleshooting. Otherwise, that network operator will have to spend a lot of time figuring out the intention of the configurations before he or she can starting any meaningful troubleshooting. This helps shorten the service-affecting event.

It can also be good to have the networking device vendor review the network configuration. The account manager of the vendor can send the configuration file back to the implementers for review and comment. While this can be tedious for the vendor, this practice can actually be mutually beneficial, because it is in the vendor's best interest to prevent any misconfiguration from causing problems in live network. The implementers may have better insight than the network operators on whether certain combinations of configurations will cause problems. For example, although configuring a large number of MPLS tunnels and configuring a large number of Ethernet Virtual LANs (VLANs) may seem independent, the MPLS tunnels and the VLANs may actually compete for some internal data structures. The system implementers may be able to spot this problem while the network operators may not.

Check for Sudden Changes

Just like a sudden change in certain part of the human body usually signals a potential health issue, a sudden change in the network (e.g., certain link's utilization) is usually a warning sign of a potential issue. For example, in a campus network, if the link utilization of each dormitory is monitored and sudden link utilization changes are recorded, many security attacks launched from that campus can be quickly detected and the perpetrators can usually be quickly identified. Similarly, in a NSP network, sudden changes in certain links' utilization usually signal a potential issue. Through auditing, network operators can spot such changes quickly. They can then try to identify the cause. In this process, the issue will likely be resolved before it causes serious problems. At the very least, such changes can be noted and correlated should a serious problem ensue.

Check for Security Loopholes

Tools can be used to launch fake security attacks and see how susceptible the network is [Arbor][Verisign]. Third-party companies can also be employed to do the security audit. This is fairly well known and practiced.

In summary, it is not easy to decide what to audit. The "how" part can be easier once the "what" part has been figured out. A rule of thumb is, after an issue has been identified and resolved, the network operators should do a post mortem to discuss whether the issue could have been prevented, and if yes, how. This may lead to the adding of an auditing procedure to prevent such issues from happening again in the future.

Another issue associated with network auditing is the lack of recognition for the auditors. People all know that prevention is more effective and less costly than curing. Ironically, preventing a problem from happening generally gets the auditor little recognition. It can be considered as "that's your job and you should do it." In contrast, the person who fixes a live network problem can get a lot of visibility and credit. If this issue is not resolved, there will be little incentive for the experts inside the NSPs to perform network auditing. NSPs would be better off setting up incentives to encourage such auditing to prevent potential problems from happening.

TRAFFIC CONTROL

Network planning sets the network structure, capacity, routing policies, and security framework. Traffic control deals with optimizing traffic distribution inside this framework, either in normal network condition or in failure scenarios. Common traffic control schemes include traffic engineering, traffic protection and restoration, and traffic management.

Again, it should be noted that certain traffic control schemes have a larger effect on network performance than others. For example, traffic management mechanisms such as queueing, scheduling, policing, and shaping can only improve performance of higher-priority traffic at a congested network device at the expense of lower-priority traffic, traffic engineering can possibly relieve the congestion at one or multiple devices, and improve performance for all classes of traffic. In other words, there are macro-control schemes and there are micro-control schemes. Depending on the scenario, one can be more appropriate than the other. Generally speaking, macro-control schemes should be considered before micro-control schemes.

Traffic Engineering

This section is focused on intra-domain traffic engineering [RFC3272]. Inter-domain traffic engineering can introduce significant traffic distribution change in a network. That is more a routing policy issue and has already been covered in the "Plan Routing Policies" section previously.

As we previously discussed, traffic engineering can be Diffserv aware [RFC3564] or Diffserv agnostic. This book advocates the use of the Diffserv-agnostic approach for three reasons:

1. A network should have sufficient capacity to ensure that every class of traffic can have sufficient capacity to avoid congestion in most cases. In the rare cases where congestion is not avoidable, Diffserv traffic management can be used to give preferential treatment to high-priority traffic. Therefore, Diffserv-aware TE is not necessary.

2. Diffserv-aware traffic engineering is fairly complex. While it can be useful in the rare cases (for example, caused by some catastrophic failures) when there is no sufficient resource for all classes of traffic, the rareness of such cases (consequence of the widely accepted ≤ 50 percent link utilization), and the absence of explicit performance guarantee (consequence of not selling CoS) make the complexity unjustifiable.

3. According to Bruce Davie, Cisco Fellow specializing on QoS, "DS-TE deployment has been impeded by a lack of edge functions to steer traffic onto the different classes of LSPs."

Intra-domain traffic engineering can be done with MPLS traffic engineering tunnels or via IGP metric manipulation. These two approaches are briefly discussed below.

Traffic Engineering with MPLS

Traffic engineering with MPLS tunnels [RFC2702] is a well-known traffic engineering approach. Many people will generally think of MPLS traffic engineering when traffic engineering is mentioned.

MPLS traffic engineering allows network operators to assign a bandwidth value to each link. The assigned bandwidth usually equals to the physical bandwidth of the link. MPLS traffic engineering also allows the assignment of a bandwidth value to each MPLS tunnel. This value usually equals to the amount of traffic carried by the tunnel, as measured. Because the assigned bandwidth values of the links and the LSPs are signaled by a protocol called RSVP (resource reservation protocol), these bandwidth values are usually called RSVP bandwidth of the links and the tunnels. When placing the tunnel onto the network, MPLS traffic engineering will ensure that on any link, the bandwidth sum of all tunnels going through that link is less than the bandwidth of the link. Conceptually, this ensures that every link has more bandwidth than needed and thus prevents congestion.

Further customizations are possible. For example, if it is desired that no link should have utilization higher than 80 percent, the RSVP bandwidth of all links can be set to 80 percent of its physical bandwidth. The links and the tunnels can also be colored such that green tunnels can only go through green links. This gives further control on how to place the tunnels onto the network, while also introduces additional complexity. Again, for any additional control, its benefit and harm (in other words, additional complexity) should be explicitly considered, and a decision should be explicitly made on whether to introduce it.

In MPLS traffic engineering, the trickiest thing is to decide the RSVP bandwidth of each tunnel. The difficulty comes from the fact that the amount of traffic carried by a tunnel varies at different times. So which value should be used? The common choices would be the peak value, the average value, and the 95th-percentile value of all measured values in a day. In the IP world, the 95th-percentile value is used for virtually all value that is time dependent. Because of the use of the 95th-percentile value, at a given time, the amount of actual traffic carried by a tunnel and its RSVP bandwidth can be very different. In most cases, the actual amount will be less than the reserved value and will not cause any congestion. However, if multiple tunnels going through a link carry more actual traffic than their RSVP bandwidth, then congestion can happen at the link. When this happens, the RSVP bandwidth of certain tunnels can be increased to match or exceed their actual traffic. This will cause the RSVP bandwidth sum of the tunnels to exceed the link RSVP bandwidth, and force some tunnels to be re-routed to other links, thus relieving congestion.

The pro of MPLS traffic engineering is that it is fairly intuitive. As long as the network has more capacity than the real demand, the tunnel routing algorithm will generally find a tunnel placement solution, which will avoid congestion. The con is the additional complexity that it introduces. First, MPLS and RSVP must be introduced. Second, the IGP used (either OSPF and IS-IS) must be augmented to support traffic engineering. Third, a full mesh of MPLS tunnels or some kind of mesh hierarchy must be introduced, which will increase network convergence time. If MPLS is also introduced for other purposes, e.g., to provide VPN service, then MPLS traffic engineering may be easier to justify. Otherwise, its benefit must be weighed against its complexity.

For an example of how to do MPLS traffic engineering in detail, readers can refer to [Xiao2].

Traffic Engineering with IGP

Before MPLS was invented, traffic engineering had been practiced for years via IGP metric-tuning. However, traffic engineering with IGP is not well known, partly because the term "traffic engineering" only became popular in the IP world after MPLS was invented. Traffic engineering is in fact the first application of MPLS. IGP metric-tuning had not been closely associated with traffic engineering from day one because it used to work on a trial-and-error basis. The amount of traffic that will be diverted from a link is not deterministic. Therefore, it is not considered as real traffic engineering, which tends to imply a more deterministic result.

Traditional IGP metric-tuning works like this: When there is congestion on a link, the IGP metric of the link is gradually increased. At some point, this will cause some shortest paths that previously went through this link to change to other links. Consequently, the amount of traffic going through that link will be reduced. The network operator will keep increasing the IGP metric until the desired amount of traffic is rerouted.

Usually, this process works pretty well, especially for networks with reasonable average utilization. Therefore, IGP metric-tuning has been the defacto intra-domain traffic engineering method for decades.

However, if there is a region inside a network that has high average link utilization (e.g., around 75 percent), then IGP metric-tuning can be very tricky. When the IGP metric of a congested link is increased, the amount of traffic diverted and where the traffic goes to are unknown to the network operator. Therefore, that traffic could go to an undesirable place and cause congestion at another link. In the worst case, when the IGP metric is increased for the newly congested link, it could cause congestion at another link. The network cannot converge to a desirable state, even though the total capacity is sufficient to carry the traffic. Even if the worst case doesn't happen, IGP metric-tuning can be tedious in a region of high link utilization, largely because of its unknown effect. In early 1999, Global Crossing faced such a situation. Because of rapid traffic growth and the delayed delivery of a critical fiber link from another carrier's carrier, its network in the San Francisco Bay Area had unusually high utilization. Tuning the IGP metric became a tedious daily job. This was why Global Crossing chose to be the first large-scale deployer of Cisco's MPLS traffic engineering solution.

Around 2000, a series of papers [Fortz1][Fortz2][Wang3] were published on how to do IGP metric-tuning in a more systematic way. In general, these methods take the traffic matrix and a network graph as input, and produce the IGP metrics of the links as output. Roughly speaking, the prescribed link metrics minimize the maximum link utilization in the network. With a simulation tool, network operators can then see how well the network will perform with such metrics and decide whether to make such changes in the network.

With such metric calculation algorithms and a simulation tool, IGP metric-tuning can become a useful traffic engineering method. This is particularly useful for optimizing the routing structure of a network in a medium time scale (for example, monthly) so that the same network can carry more traffic without increasing the maximum link utilization.

One may be concerned that changing IGP metrics of many links may be tedious and cause some network instability. This is partly true. While a script may enable a network operator to change the IGP metrics of a large number of links in less than 10 minutes so that this can be done in a maintenance window during some midnight hour, such changes will cause a lot of routing recalculation and may affect the stability of the control plane. Therefore, the improved IGP metric-tuning method is suitable more for strategic (that is, mid/long-time scale) traffic engineering than for tactical traffic engineering (that is, real time to fix a congestion problem). But arguably, one can use the improved IGP tuning for mid/long-time scale traffic engineering and use the traditional IGP tuning for fixing congestion in real time. If a NSP is not using MPLS for purposes other than traffic engineering, this may become an option.

Traffic Protection and Restoration

Traffic protection and restoration is to detect failure quickly and switch traffic to alternative paths so that as few packets are dropped as possible. The common protection and restoration approaches are:

- MPLS fast reroute
- MPLS switchover with active/standby LSPs
- IGP fast convergence

The restoration speeds of these approaches are slightly different. For MPLS Fast Reroute, it is around 50 ms; for MPLS failover, 100 ms to several 100 ms depending on the failure-detection mechanisms; and for IGP fast convergence, sub-second.

Benchmark time for failure detection and switchover is 50 ms in the industry. This number came from the PSTN world. Today, many people believe that IP networks also need to provide 50 ms failure detection and switchover, possibly for the following reasons:

- PSTN and optical transport networks can do it, and networks are converging to IP, so IP networks need to do the same.
- Premium applications running over IP, for example, TV, require it.
- With MPLS Fast Reroute [RFC4090] or some other mechanisms such as [BFD], IP networks can do it.

Consequently, MPLS Fast Reroute became the most talked about scheme for protection and restoration.

While it is desirable for IP networks to match PSTN and optical transport networks to provide 50 ms failure detection and switchover, it is much more complicated to do this in IP than in PSTN/optical transport. The reason is, in the IP case, the decision must be made for millions of packets per second while in the TDM case, the decision needs only be made for thousands of time slots per second. Therefore, the tradeoff between benefit and complexity must be made again.

To make the subsequent discussion more meaningful, let's first explain the complexity and issues associated with 50 ms failure detection and switchover in IP/MPLS. This will make it easier to make the tradeoff between the benefit and complexity.

First, MPLS Fast Reroute involves a large amount of complexity, as explained in the "Complexity Challenge" of Chapter 8. To briefly recap here, to fast reroute a LSP, every node except the destination must set up a detour LSP to protect against failure of the next node. This increases the number of LSPs in the system, and thus the signaling complexity and convergence time. Although facility backup can allow different LSPs to share their detours, this creates coupling among LSPs, and could reduce stability of the network.

Second, if the IP/MPLS layer failure detection and switchover take effect around 50 ms like the optical layer, the interaction between IP layer protection and optical layer protection becomes a little unpredictable. The reason is, if there is a failure, both layers can detect it at about the same time and take action at about the same time. So the IP/MPLS layer may reroute the LSP to other links. However, after the optical layer protection takes effect, the L2 link layer will remain intact despite the failure at the physical layer. So from the IP layer's perspective, no failure should have been detected in the first place. The rerouted LSP may have to be restored back again. This creates unnecessary change in the network and will increase packet jitter and possibly maximum delay. From this perspective, 50 ms or below 50 ms IP/MPLS layer failure detection and switchover may not even be desirable, if optical layer protection is present.

This issue has been known since the early days of MPLS Fast Reroute. However, little progress has been made in addressing this. In 2007, there was a special issue in *IEEE Journal of Selected Areas in Communications* [JSAC] on this topic. However, no good solution was proposed. One possibility for the lack of progress is IP/MPLS Fast Reroute has not been widely deployed despite its seeming desirability. Therefore the issue is just a potential one, not a real one.

Third, for L2 Ethernet services, which involves MAC learning and broadcast of unknown MACs, 50 ms detection and switchover can be difficult to achieve. To understand why it's easier to achieve 50 ms switchover for L3 services than for L2 services, one needs to understand that the forwarding architecture for IP prefixes is generally different from that for MACs. For example, in the Forwarding Information Base, an IP prefix generally points to an "interface table," which again points to an "outgoing port table." In contrast, a MAC entry generally points directly to an "outgoing port table." While this is not easy to understand, the conclusion is that because of the additional level of indirection, there are fewer table entries to update to enable Fast Reroute in the IP forwarding case than in the MAC forwarding case. This is especially true when broadcast or multicast is involved. At least for broadcast/multicast traffic, the switchover time will likely be around 150 ms or more. Generally, the people with intimate knowledge about product implementation know this. However, the general marketing people may not be aware, or may feel pressured to match competitors' 50 ms claim.

Consequently, the 50 ms number is frequently mentioned in the industry although in reality, 50 ms may not be achievable, at least not for broadcast and multicast L2 traffic. Therefore, the benefit of MPLS Fast Reroute needs to be quantified as well, depending on the services being protected.

After discussing the complexity and issues associated with MPLS Fast Reroute, the next question would be "is it a must-have?" If it is, then NSPs must deploy it despite the complexity and issues. That depends on the answers to the following 2 questions:

1. Do applications require 50 ms detection and switchover?
2. Must it be replicated at the IP layer or can it possibly be done at a different layer?

For the purpose of this discussion, let's assume that the answer to the first question is yes. Otherwise, MPLS Fast Reroute won't be a must-have. But still, the answer to the second question will not be a clear "yes," because 50 ms protection can possibly be done at a different layer. The reason is, while most applications will probably run over IP in the future, this is regarding the end points only. For the network in the middle, not every packet needs to go through IP. For example, a video server and a video player can both be based on IP. However, they can be connected by a wire without IP (or any other protocol). For a NSP that is providing triple (or quadruple) play services, different services can be delivered at different layers of the network. For example, real-time broadcast TV (BTV) can be delivered very efficiently with a feature called drop-and-continue. That is, the TV signals are dropped at the first place, and continued to the second place, dropped there and continued to the third place, and so on. As of May 2007, the world's largest [IPTV] deployment was by [PCCW], the incumbent NSP in Hong Kong. PCCW uses this approach. Drop and continue can be done at either the optical layer or with [RPR] or possibly other technology, which is usually in some layer below MPLS.[1] Similarly, the largest IPTV deployment in the United States (by Verizon) delivers IPTV optically via the downstream broadcast channel in GPON. The 50 ms protection and restoration for BTV can therefore be provided by the optical layer. 50 ms protection in IP/MPLS may not be necessary.

Similarly, VoIP and VoD services can be protected at the optical layer if applicable.

The discussion above does not mean to say that MPLS Fast Reroute should not be deployed. In fact, in recent years, a few NSPs deployed MPLS Fast Reroute to provide fast protection for VoIP. The discussion here just points out MPLS Fast Reroute has its complexity and issues, and 50 ms protection and switchover can possibly be done at a different layer. Therefore, network operators should make the explicit tradeoff between benefit and complexity to decide whether to deploy MPLS Fast Reroute.

---

[1] A significant advantage of such an approach is multicast can be eliminated from the IP/MPLS layer. This simplifies network operations significantly because multicast is difficult to manage [McastOps].

Beside MPLS Fast Reroute, MPLS switchover with active/standby LSPs and IGP fast convergence are two other protection approaches. They are discussed below.

With MPLS switchover, the head node of a LSP will also set up a backup LSP to the destination. In normal conditions, packets are only forwarded on the active LSP. When the active LSP fails, packets will be sent over the backup LSP. Originally, the detection of failure of the active LSP is based on RSVP-TE Hello messages. For example, if a number of Hello messages are missed, then the active LSP will be considered down. Packets will then be switched to the standby LSP.

Lately, Bi-directional Forwarding Detection (BFD) has been proposed as the mechanism for detecting the failure of the active LSP [BFD]. BFD involves less processing and therefore can be sent more frequently. This enables the head node to detect failure and switchover faster. Depending on the BFD timer, this generally works in the 100 ms time scale. Using BFD as a failure detection mechanism, MPLS switchover can take effect in the 100 to 250 ms time scale.

IGP generally converges in the range of seconds. In 2001, Van Jacobson et al. proposed that IGP Hello timers be reduced to bring the convergence time down to a sub-second [IGPfast]. After BFD was invented, some people also proposed using BFD as the failure detection mechanism because of its lower processing overhead. Fast IGP convergence generally takes effect in sub second.

One can see that MPLS switchover and IGP convergence are relatively simple compared to MPLS Fast Reroute. They take effect in the 100 ms range so they won't interfere with optical layer protection either. They should be considered before one resorts to MPLS Fast Reroute.

Another interesting point to note is, despite common belief, IP devices are not that reliable as PSTN devices, largely because their complexity is much higher [RFC3439]. Therefore it would be more productive to reduce complexity to increase device reliability, instead of adding further complexity to provide protection and restoration after a device failures.

Traffic Management

In this book, we advocate not selling CoS. That doesn't mean that we advocate not using Diffserv or traffic management at all. At needy times, NSPs can use CoS and traffic management to improve service quality in their networks. For example, if certain links are congested because some failures have happened and other traffic control schemes such as traffic engineering have not been able to prevent congestion, traffic management can be used to provide preferential packet treatment to applications that are sensitive to loss, delay, and jitter. In other words, traffic management can be used to supplement other traffic control schemes, as a last resort to handle congestion. Better service quality resulting from that can become a competitive advantage.

As described in Chapter 4, common traffic management mechanisms include:

- Classification and marking
- Policing and shaping

- Class-based queueing and scheduling
- Weighted Random Early Detection (WRED)

If traffic management is deemed necessary, the traffic management approach described in Chapter 4, as used by the traditional QoS approach, can be used. It won't be repeated here. The following suggestions are provided for the use of traffic management:

- Activate Diffserv and traffic management only at interfaces that are congested.

- At interfaces where congestion is possible, prepare the Diffserv and traffic management configuration ahead of time so that they can be quickly enabled when congestion happens.

TRAFFIC OPTIMIZATION

CDN

Content Delivery Network (CDN) essentially replicates content to multiple places so that content becomes closer to the end users. The content replication can be proactive or passive. With proactive replication, content is replicated to servers at different places before it is requested. Many commercial CDN companies take this approach [Akamai][Internap][Savvis]. With passive replication, content is replicated (for example, cached) to a certain server(s) after it is used. To a certain extent, this is a tradeoff between cost and performance.

CDN simultaneously improves two important QoS factors. First, by replicating hot content to multiple servers, CDN prevents servers from becoming overloaded. This enables fast server response. Second, by moving content closer to the end users, the number of intermediate devices that can introduce packet loss, delay, and jitter is reduced, resulting in smaller packet loss, delay, and likely jitter. Therefore, CDN can be very helpful in improving end users' QoS perception. In fact, they can be more effective than traffic control schemes such as traffic engineering, traffic protection and restoration, and traffic management. Because of that, many major online content providers use some sort of CDN service provided by [Akamai][Internap][Savvis].

Because this chapter is mainly concerned with how to provide QoS in a network, so a natural question is: Should NSPs themselves strive to provide CDN service? Because CDN is more about deploying a large number of servers close to the end users and replicating content among these servers, and less about the network infrastructure itself, NSPs that own network infrastructure do not necessarily have any inherent advantage over the traditional CDN providers that concentrate on server deployment and content replication. Therefore, it may be advisable for NSPs to focus on managing their own network infrastructure, and partner with CDN providers to deliver good QoS to their end users.

Route Control

Route control is to pick the best-performing Internet path to deliver customers' traffic. As described previously in this chapter, for two external destinations with the same WEIGHT and LOCAL_PREF, the path with the shortest Autonomous System Path (AS_PATH) will be used. This is similar to using the shortest IGP path inside a domain. It generally makes sense. However, just like a shortest IGP path can be congested, a shortest AS_PATH can have some performance issues at times. Therefore, if the performance of the various paths to the same destination AS is measured, then a non-shortest AS_PATH path can be used at times to provide better performance [Internap].

To a certain extent, route control is an inter-domain traffic engineering scheme. As discussed previously, inter-domain traffic engineering can introduce significant traffic shift in one's network. Therefore route control is suitable for "thin" NSPs. Such a NSP is just a thin layer between its customers and other NSPs. As soon as it gets traffic from its customers, it will pick the next NSP to send the traffic to based on a certain performance metric. Therefore, the significant traffic shift will only involve a few devices, and will be manageable. In contrast, NSPs with a large network will not benefit as much from route control. This is because many more devices will be involved in the significant traffic shift, and the overhead of managing such a shift will be higher.

PERFORMANCE MEASUREMENT

"You get what you measure" is an old saying in the business world. It means that if a business wants to excel in certain aspects, for example, superior customer service, then it should measure its performance in those aspects quantitatively. This is also applicable in the delivery of QoS. After careful network planning, diligent network auditing, and possibly some traffic engineering, networks should stay in good condition most of the time. Satisfactory QoS should result. When some failure happens, various traffic control schemes should take effect and hopefully prevent or relieve congestion. Satisfactory QoS should still be maintained. But there may still be some unexpected factors, e.g., human errors or security loopholes, causing the network not to deliver the performance as expected. Therefore, it is only through meticulous performance measurement that one can be sure of the exact performance of one's network. Performance statistics will also provide a direction of what can be improved, and provide input for the next round of network planning. This thus forms a complete control loop.

Statistics Collection

For statistics collection, the two key questions are what and how. In terms of what, the key performance statistics that should be measured are delay, delay variation, and packet loss ratio. These statistics can determine the performance of end users' applications. In terms of how, NSPs usually install some devices in their PoPs for such a purpose. These devices could be general purpose PCs, e.g., running Linux Operating System, or some customized performance measurement devices. Generally, these devices use ICMP ping to determine the delay, delay variation, and packet loss ratio to measurement devices in other PoPs. This takes care of the network part of performance measurement. For the network-customer part of performance measurement, with the consent of a customer, the NSP's measurement device can ping a device of the customer, e.g., the CPE router, to determine the performance of the portion of the network between the customer and the NSP. However, this approach has certain drawbacks. For example, the customer's CPE router will be susceptible to external ICMP attacks. Consequently, some people believe that some special purpose performance measurement devices are needed at the customer's premises.

Statistics Analysis

Performance statistics collected by the measurement devices are exported to a central location regularly. The statistics can then be analyzed. The main goals of the analysis are, first, to find out whether the performance meets the need of end users' applications; and second, to find out whether there is anything unusual. The first goal is relatively easy to accomplish, as the requirements of the end users' applications are fairly well defined. A simple comparison would give the answer. The second goal is not as well defined, as "what is unusual" is not always clear. Therefore this is similar to network auditing and the discussions there are applicable here. Tools can help in this regard, for example, [WANDL] and [Cariden].

CONTROL SCHEMES THAT ARE NOT RECOMMENDED

In this section, we discuss a few schemes that are frequently mentioned in the area of QoS. These are the schemes that, in the author's view, are too complicated to justify. Although these schemes can theoretically be useful if network utilization is unusually high (e.g., average utilization exceeding 70 percent), in reality such cases are very rare. Therefore, they are not recommended.

RACS

Resource and Admission Control Subsystem (RACS) is defined by the Telecoms & Internet converged Services & Protocols for Advanced Networks ([TISPAN]). TISPAN itself is a standardization body of European Telecommunications Standards Institute (ETSI). Conceptually, RACS works as follows.

When a user needs priority treatment for a traffic stream, it would send a request for a resource to the RACS server of the local domain, which maintains resource utilization information of the network. Depending on whether the resource is available, the RACS server will either grant or deny the request. Because the priority treatment needs be done end to end, the RACS server of a domain will likely have to consult with RACS servers in other domains before granting a request. If a resource request is granted, then the RACS server will also inform the network devices along the path of the traffic stream to reserve the resource for the traffic stream. In other words, RACS works very much like the Resource Reservation Protocol (RSVP) as defined in [RFC2205]. The advocators of RACS may point out that RACS is different from RSVP in that it is integrated with accounting and billing support.

To provide hard QoS guarantee for a traffic stream, theoretically a resource must be reserved for that stream. Otherwise, there is always a possibility that the resource is used by other traffic and the QoS guarantee of that stream can't be met. This is also very similar to how NSPs provide toll quality to voice. Therefore, RACS (or previously RSVP) may appear appealing. However, in reality it may not be useful for two reasons.

First, unlike the PSTN days, today's capacity is at a much lower cost. Because of this and IP traffic's fast growth and perceived burstiness, IP networks are generally provisioned with more idle capacity [Odlyzko2]. Furthermore, researches in recent years indicated that for links with speed higher than 50 Mbps, IP traffic is not as bursty as people expect [Fraleigh]. The reason is at high speed, there will be sufficient aggregation to reduce the bursty effect. Therefore, IP networks generally have more headroom to handle bursty requests. This reduces the need for resource reservation.

Second, RACS cannot really provide hard guarantee on QoS after other factors such as device failure and operator errors are considered. This becomes apparent when reliability is considered as an important factor for QoS, as we advocate here. It is not so in the traditional QoS wisdom where QoS and reliability are considered independent. When all factors are considered, it is not clear that it will be useful to seek perfection in one QoS factor, i.e., to guarantee that the resource will be available for the admitted traffic streams, while introducing significant complexity that will compromise other QoS factors (e.g. device reliability, configuration auditing, etc). This is especially true given that the Internet in developed countries is already good enough for premium applications, and the problems are mostly related to failure, which is primarily associated with complexity. The evidences for the last statement were provided in Chapter 2, and in the "Network Auditing" section of this chapter.

Just like RSVP as a mechanism for providing QoS is never deployed, it is unlikely that RACS will have any sizeable deployment. The fact that RACS has an advantage over RSVP in that it provides accounting and billing of the requested resource may not matter, because accounting and billing only matter after RACS itself is justified.

Diffserv-aware TE

Diffserv-aware TE and its complexity have been discussed in Chapter 8. We caution against Diffserv-aware TE because of its high complexity. Diffserv-aware TE can only be justified if a NSP requires both Diffserv and TE. Even that is just a necessary condition, not a sufficient one, for Diffserv-aware TE to be considered. When only 10 percent of the world's NSPs have deployed Diffserv-agnostic TE (as of Oct. 2007[2]), and no NSP has generated much revenue from Diffserv, the benefit of Diffserv-aware TE does not seem to justify its complexity.

DIFFERENCES FROM THE TRADITIONAL QOS APPROACH

Here we briefly summarize the differences between the proposed technical approach for providing QoS and the traditional traffic management-centric approach. The major differences are:

- Congestion prevention is emphasized over congestion handling.
- Various traffic controls are prioritized and integrated.
- Tradeoff between the benefit of traffic controls and their complexity is explicitly made.

In the proposed technical approach, "preventing disease" is preferred over "curing disease." Therefore network planning is emphasized to make sure that the network has sufficient capacity in normal conditions and is robust against common failures. Network auditing is also emphasized to reduce human errors and their detrimental effect. Consequently, networks should become "healthier," and when they have issues, "cures" should be simpler. In other words, this reduces the need for complicated traffic control schemes in the network.

Of the various traffic control schemes such as traffic engineering, traffic management, and traffic optimization, their effectiveness for QoS is prioritized. Again, congestion avoidance and relief are emphasized over congestion handling. Therefore, macro-control schemes such as traffic optimization and traffic engineering are emphasized over micro-control schemes such as traffic management. This addresses the integration challenge of the traditional QoS approach.

In the proposed technical approach, the complexity of each traffic control scheme is explicitly considered, and traded off against its benefit. This avoids the introduction of too many controls into the network and prevents the control/complexity spiral from happening. This addresses the complexity challenges of the traditional QoS approach.

BENEFITS OF THE PROPOSED APPROACH

The benefits of the proposed approach are simplified network operations and simplified network equipment.

---

[2] This is according to WANDL, a company that provides MPLS-TE design and simulation tools to NSPs. Those NSPs that deployed TE are usually the incumbent NSPs that have more talents.

Simplified Network Operations

In the proposed technical approach, by emphasizing network planning and network auditing, networks will have fewer issues and will rely less on complicated traffic control schemes in handling issues when they do happen. Also, by explicitly trading off complexity and benefit of each traffic control scheme, fewer control schemes will be introduced into the networks. For example, with the proposed approach, network planning should limit the need of Diffserv traffic management to only a few network devices, network planning and simple rate limiting may remove the need for RACS, IGP traffic tuning and network simulation tools may remove the need for MPLS-TE, optical protection or active/standby LSPs with BFD may remove the need for MPLS Fast Reroute, and Diffserv-agnostic TE may remove the need for Diffserv-aware TE. Consequently, networks will be simpler to operate. This will lead to lower OPEX.

As the proposed technical approach is focused on providing good performance for all classes of traffic, not on providing perceivable differentiation among different traffic classes, it also simplifies the interoperation with other NSPs. This is because providing differentiation end to end for different traffic classes requires careful coordination among NSPs, while providing best possible services requires far less coordination beyond what is already done today. This will also lead to lower OPEX.

Simplified Network Equipment

When NSPs rely less on complicated traffic control schemes, network equipment vendors will have less pressure to provide such complicated schemes in their devices. The incumbent vendors may still want to advocate such complicated schemes to create entry barriers for other vendors. But if the industry is consciously aware of the tradeoff between the benefit of controls and their complexity, then the advocators of complicated schemes will have less success. Consequently, future network equipment may become simpler if the proposed technical approach for QoS prevails. For example, RACS and Diffserv-aware TE are both very complicated software to implement. Not having to implement them in future network devices would save network equipment vendors a significant amount of resources. Similarly, by de-emphasizing differentiation among different traffic classes and emphasizing providing best possible service for all classes, the need for a large number of traffic queues can be reduced. This is because queues are primarily used for separating out different traffic so that they can be treated differently. This will lead to saving of hardware resources, e.g., memory for queues and hardware logic to manipulate the queues. All of these will lead to higher reliability and lower cost of network equipment.

THE EARLY EVIDENCE

Compared to the traditional QoS approach, which focuses on congestion handling, the proposed approach emphasizes congestion avoidance. However, when there is congestion, the proposed approach allows the quick invocation of Diffserv traffic management for congestion handling. Therefore, there should be no doubt that this approach can deliver the QoS better than the traditional QoS approach.

What we will show in this section is, although existing QoS literature gives the impression that Diffserv traffic management is the most important way of delivering QoS, the real world has largely succeeded in using capacity (and sound networking practice) to deliver QoS for premium applications without using much Diffserv traffic management. In other words, the proposed approach is already in use by many real-world NSPs with considerable success.

First, to briefly recap the discussion in Chapter 2, today's Internet in developed countries is already providing good performance on delay, delay variation, and packet loss ratio. More specifically:

- For delay: round-trip delay measured from Stanford University to most countries in the world except some in Africa and Central Asia is below 250 ms [ICFA07]. This means that within each region, especially within North America, West Europe, and East Asia (China, Japan, Korea), one-way network delay is likely around or below 100 ms. This largely meets the delay requirements of premium applications;

- For delay variation: from some fairly realistic calculations done in Chapter 2, it is shown that delay variation can largely be bounded within 50 ms, especially within developed regions and with the use of jitter buffer. This largely meets the delay variation requirements of premium applications;

- For packet loss ratio: packet loss ratio measured from Stanford University to developed regions such as North America, West Europe, and East Asia is at 0.1 percent level [ICFA07]. Therefore, within these regions, there is a good possibility that the actual packet loss ratio will be below 0.1 percent. Granted that the reported loss statistics are monthly statistics and instantaneous loss can be higher at times, this nonetheless means that in developed regions, packet loss ratio performance meets the requirements of premium applications in normal network condition.

Second, there is strong evidence that the Internet's performance is improving considerably year after year. For example, Figure 11-5 shows that the Round-Trip Time (RTT) performance from the United States to many countries (most notably China and India) improved from "bad" to "acceptable" from 2000 to 2003.

Figure 11-6 shows the improvement of RTT over time in a more quantitative way. First, every region except Central Asia is improving. Second, since 2006, round-trip delay has become acceptable in other regions except Africa and Central Asia.

Figure 11-7 shows the improvement of packet loss ratio over time. By October 2006, packet loss ratios measured from Stanford University to North America, Europe (except the Balkans and Russia), East Asia, and Oceania are all below 0.1 percent. More importantly, this figure shows that packet loss ratio performance

FIGURE 11-5

Minimum RTT improvement from 2000 to 2003

If the Legend is not clear in black and white, readers are encouraged to check out the color picture in [ICFA07] for more clarity.

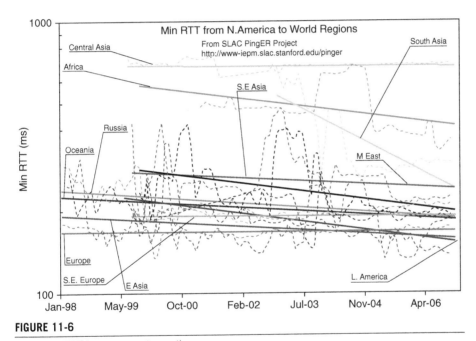

FIGURE 11-6

Minimum RTT improvement over time

is improving very fast year after year. Of the three major performance criteria, network delay and delay variation already (or come very close to) meet the requirements of premium applications. Packet loss ratio also meets or comes close to meeting the requirements but is nonetheless the biggest remaining challenge. Therefore, significant improvement in packet loss ratio year after year is very encouraging.

In general, [ICFA07] concluded that Internet performance is improving each year, with throughputs typically improving by 40–50 percent per year and losses by 25–45 percent per year. Therefore, even if it's not good enough somewhere

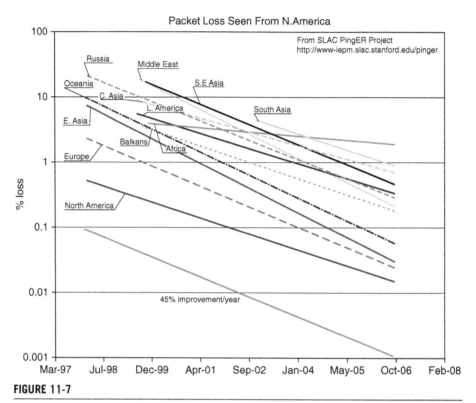

FIGURE 11-7

Packet loss ratio improvement over time

today, it will likely become acceptable at some point in the future. And all of these are done without much active traffic control.

SUMMARY

The key points of this chapter are:

- Network planning is essential for providing QoS. By making sure that networks have sufficient capacity in normal condition and are robust against common failures, networks will experience fewer congestions and will rely less on complex control schemes. Both will lead to better QoS;

- Network auditing is also essential for providing QoS. By trying to reduce human errors and their detrimental effect, networks will have fewer issues and thus better QoS;

- Traffic control schemes such as traffic optimization, traffic engineering, and traffic management have different effectiveness on QoS. The order they are applied should be considered. In general, it is recommended that macro-control schemes such as traffic optimization and traffic engineering be considered before micro-control schemes such as traffic management;

- Network performance should be measured to find out the exact performance of the network and whether anything should/can be improved. Performance measurement completes the control loop for the delivering of QoS;

- The major differences between the proposed technical approach and the traditional traffic management centric are: First, congestion prevention is preferred over congestion handling; second, various traffic controls are prioritized and integrated; and third, tradeoff between the benefit of traffic controls and their complexity is explicitly made;

- The benefits of the proposed approach are simplified network operations and simplified network equipment;

- There is early evidence that the proposed approach would work to deliver the needed QoS for applications.

Up to this point, we've discussed the challenges of the current QoS model, and proposed a revised business model and a revised technical solution. In the following chapters, we will present two case studies of real-world QoS deployment, and discuss QoS issues in wireless networks.

Please voice your opinion about this chapter at http://groups.google.com/group/qos-challenges.

Case Studies

12

This chapter is written by outside contributors. The purpose is to give the readers an opportunity to hear directly from the horse's mouth regarding real-world QoS deployment. The author just provided the desired organization for each case study. The content is provided by the contributors. The themes of the case studies are consistent with the theme of this book. However, certain specific opinions differ. This is considered a good thing: Different opinions can give the readers a diversified view on QoS, and help the readers form their own opinion.

In the first case study, Ben Teitelbaum and Stanislav Shalunov share their experience of trying to use Diffserv to provide a "Premium Service" for the Internet2 community, the lessons they learned, and the conclusions they drew. Internet2 is the backbone network for the education and research institutes in the United States. Ben Teitelbaum and Stanislav Shalunov were the architects of Internet2 from 1997 to 2006 before they moved to BitTorrent. "Premium" is a better-than-Best-Effort service that will make an Internet connection behave like a hard wire. The original QoS approach that they took for Internet2 was a typical traffic-management-centric approach. Therefore, the lessons they learned and the conclusions they drew can be very useful for people who are accustomed to the traditional QoS approach. Of course, it is worth noting that Internet2 is a backbone network. Therefore the lessons and conclusions may not be directly applicable to access and aggregation networks.

In the second case study, Ricky Duman presents how Internap uses route control and CDN to provide QoS. This is a non-traditional approach for QoS, but with it, Internap is able to offer one of the most aggressive SLAs in the industry in terms of delay, delay variation, packet loss ratio, and network availability. Internap is one of the few companies that successfully generated revenue from QoS without relying on CoS. Therefore, their experience and conclusions offer a fresh perspective on QoS.

The contributions of Ben Teitelbaum, Stanislav Shalunov, and Ricky Duman are gratefully acknowledged.

CASE STUDY 1: DELIVERING QOS AT INTERNET2 (BY BEN TEITELBAUM AND STANISLAV SHALUNOV)

Ben Teitelbaum
BitTorrent, Inc. ben@teitelbaum.us

Stanislav Shalunov
BitTorrent, Inc. shalunov@shlang.com

Introduction to Internet2

Internet2 is the foremost U.S. advanced networking consortium. Established in 1996, Internet2 facilitates the development, deployment, and use of revolutionary Internet technologies. Internet2 includes campus, regional, and nationwide networks. It connects over 5 million individuals at more than 240 research and education institutions in the United States. It also interconnects with research networks in dozens of other countries. The Internet2 IP Network supports IPv4, IPv6, multicast, and other advanced networking protocols. The diagram of Internet2 is shown in Figure 12-1. Note that "connectors" are the member organizations that are directly connected to the Internet2 network. Other organizations may be connected to the Internet2 IP Network via the connectors. Each connector is a domain.

Internet2 charges its member organization a participation fee, which is $22,000 for 2007 and $24,000 for 2008. In addition, there is a network connection fee as described in Table 12-1. More information can be found at: http://www.internet2.edu/network/fees.html.

FIGURE 12-1

Network diagram of Internet2

Table 12-1 Internet2 connection fee at various link speeds	
Connection type	**Connection fee for 2007**
OC-3	$110,000
OC-12	$220,000
GE	$250,000
2 × GE One link provides connection to IP services, the other to the wavelength switching network as a backup link	$340,000
OC-48	$340,000
10GE	$480,000
2 × 10GE One link provides connection to IP services, the other to the wavelength switching network as a backup link	$550,000

Original QoS Deployment Plan of Internet2

At its inception, one of the primary technical objectives of Internet2 was to engineer scalable, interoperable, and administrable inter-domain Quality of Service (QoS) to support an evolving set of new advanced networked applications. It was envisioned that QoS technologies would prioritize certain applications, such as streaming video or videoconferencing, to assure that their bits traversed the network with the required reliability and fine-tuned performance. To achieve this goal, in the fall of 1997, the Internet2 QoS Working Group worked with advanced applications developers and network operators, and identified a demanding set of requirements for Internet2 QoS. Foremost among these requirements was the need to meet the absolute, end-to-end performance requirements of a broad range of advanced applications, including several important applications requiring hard real-time assurances. From Internet2 network engineers and planners came two additional technical requirements: (1) any viable approach must scale, allowing core routers to support thousands of QoS-enabled flows at high forwarding rates, and (2) any viable approach must interoperate, making it possible to get well-defined inter-domain QoS assurances by concatenating the QoS capabilities of several independently configured and administered network clouds. As the Internet2 QoS Working Group looked for solutions to these problems, the Diffserv approach [RFC2474][RFC2475] began to gain significant interest in the IETF as a lightweight alternative to the Integrated Services (IntServ) approach [RFC1633]. In May 1998 the working group presented Diffserv as the architecture best suited to meeting the QoS needs of Internet2. The [QBone] initiative, launched in late 1998, aimed to build an open and highly instrumented testbed for inter-domain differentiated services. Networks participating in the QBone included vBNS, Abilene,

ESNet, NREN, CA*Net2, SURFnet, TransPac, MREN, NYSERNET, NCNI, the Texas gigaPoP, and numerous universities and labs. QBone focused initially on deploying and evaluating a single service, the QBone Premium Service (QPS). QPS aimed at making an IP connection behave like a hard wire, i.e., with deterministic delay, jitter, and no packet loss. QPS was based on the Expedited Forwarding (EF) [RFC2598].

The planned QBone architecture is described below.

Overview

The QBone architecture sought to build an inter-domain service, an integrated measurement collection and dissemination architecture, and a set of common operational practices for establishing inter-domain reservations. The architecture was expected to evolve incrementally as experience with the QBone testbed was gained.

QBone Premium Service

The QBone Premium Service (QPS) would make inter-domain, peak-limited bandwidth assurances with virtually no loss, and with virtually no delay or jitter due to queuing effects. The intent of the QPS service was to approximate the "virtual leased line" or "Premium" service proposed by Van Jacobson in [2Bit] and initially demonstrated across Abilene, ESNet, and the LBNL QoS testbed to the show floor of SuperComputing 2000 [CBQdemo]. QPS exploited the Expedited Forwarding (EF) per hop forwarding behavior. EF requires that the "departure rate of the aggregate's packets from any Diffserv node must equal or exceed a configurable rate" and that the EF traffic "SHOULD receive this rate independent of the intensity of any other traffic attempting to transit the node." EF may be implemented by any of a variety of queueing disciplines, but is best thought of in terms of forwarding EF packets through a strict priority queue. Services like QPS were built from EF through careful conditioning of EF aggregates so that the arrival rate of EF packets at any node was always less than that node's configured minimum departure rate. A QPS reservation {source, destination, route, startTime, stopTime, peakRate, serviceMTU} was an agreement to provide the transmission assurances of the QBone Premium Service starting at startTime and ending at endTime across the domain-to-domain chain route between source and destination (these are arbitrary network prefixes) for EF traffic entering at source and conforming to the token bucket traffic profile parameterized by:

- token rate equal to peakRate bytes per second; and
- bucket depth equal to serviceMTU bytes.

The transmission assurance offered by the QBone Premium Service is as follows:

- Low loss: this should be very close to zero, but will not be quantified in the service definition.

- Low latency: queueing delay will be minimized, but no assumption regarding minimal latency was made.

■ Low jitter: delay variation caused by queueing effects should be no greater than the packet transmission time of a service MTU-sized packet at the subscribed rate; no assumption was made about jitter due to other effects (for example, route instability).

Traffic exceeding the profile {peakRate, serviceMTU} would be dropped on ingress and not allowed to progress into downstream DS domains. Bilateral QPS reservations had the same structure as wide-area inter-domain QPS reservations and had comparable implications for the configuration of ingress traffic conditioners. Note that if two domains wanted to use different serviceMTUs, then reshaping must happen at the boundary if going from a larger to a smaller value. Consistent with the Diffserv architectural model, all QPS SLAs were determined bilaterally between adjacent QBone networks (dubbed "DS domains"). Although the agreements that defined SLAs were strictly bilateral, there were technical implications of the QPS described above that impose minimum requirements on QBone SLAs. These requirements, as well as certain recommendations regarding the EF codepoint and the routing of EF traffic, were discussed in the QBone Architecture document [QBarch].

Measurement architecture

To debug, audit, and study QBone services, each participating QBone domain must collect and disseminate a basic set of QoS measurements. The QBone architecture specified the measurement metrics to be collected at the edges of each participating QBone domain, and specified how these data were to be disseminated to the community. It was expected that the QBone measurement architecture would evolve with the QBone itself, as new measurement metrics and reporting techniques were modified or added with experience.

The metrics required of each QBone domain may be obtained through a combination of techniques, including active measurement, passive "sniffing," and SNMP-style polling of network devices. All measurements were to be taken at or as close to inter-domain boundary routers as possible. Each edge router of a QBone domain must be instrumented to serve as a "QBone Measurement Node." When possible, active measurement probes would have direct interfaces to boundary routers. These probes were the sources and sinks of dedicated measurement paths terminating at the measurement node. Passive measurement probes must observe the traffic on inter-domain links without perturbing it in any respect. This was commonly accomplished through the use of an optical splitter. Passive measurement probes may additionally be located on intra-domain links to assist in deriving metrics such as expedited-forwarding and Best-Effort interface discards. Passive measurement equipment may also be used to measure probe flows created by active measurement equipment. SNMP-style polling agents that extracted router statistics to support QBone utilization metrics may be physically located anywhere. In addition to specifying the metrics that must be collected by each participating QBone domain, the QBone architecture specified the paths (defined by ordered pairs of measurement nodes) along which path-based metrics, such as one-way packet delay variation, are to be collected.

The initial design of the QBone measurement architecture was focused on verifying the service goals of the QBone Premium Service and helping with the debugging, provisioning, and performance of EF behavior aggregates. In specifying which metrics were required by the QBone architecture, a concerted attempt was made to strike a balance between an ideal of what would be interesting to measure and a real-world pragmatism that considered only those metrics for which cost-effective implementation techniques were known to exist. Although the QBone architecture made no specific implementation recommendations for the collection of the required measurement metrics, a companion document was then under construction that would provide this guidance to QBone participants. The required metrics were described below. For each metric, data were to be collected for both the EF and Best-Effort (BE) behavior aggregates. The first class of required metrics was those that must be obtained through active measurement of paths between QBone boundary nodes. The purpose of these measurements was to build a picture of the network's behavior continuously over time. These measurements will add a small background load to the traffic in each behavior aggregate, but relative to the capacity of the links along each test path, this should be negligible. Small EF aggregates must be reserved for EF active measurement streams, which must take care not to exceed the service parameters of these reservations. Required active metrics for each test path include:

- the IETF one-way packet loss metric [RFC2680];
- an instantaneous one-way packet delay variation metric based on [RFC3393];
- periodic traceroutes for each behavior aggregate (formal metrics such as for inter-domain path stability could later be derived from these).

The second class of required metrics must be obtained passively—either through sniffing of inter-domain links at QBone boundary nodes or through SNMP polling of edge devices for key interface statistics. There was only one metric in this class:

- EF and BE load measured in packets per second and bits per second.

A final class of required metrics must be obtained through polling of (largely static) provisioning data and through polling of each QBone domain's reservation system. These metrics included:

- link bandwidth in IP bits per second;
- EF commitment in IP bits per second (this was the maximum EF load that a QBone domain would be willing to carry over the given link);
- EF reservation load in IP bits per second (this was the QPS peakRate configured for the given link).

In addition to the required metrics discussed above, the QBone architecture recommended the optional collection of several metrics that were either difficult to collect or were not yet precisely defined. These metrics were suggested because their collection would add significant value to the QBone measurement infrastructure. Suggested metrics included: EF and BE interface discards, one-way packet delay, and end-to-end burst throughput tests.

A key goal of the QBone measurement architecture was to collect, disseminate, and present results in a consistent fashion. This uniformity would greatly simplify the analysis of inter-domain QPS reservations and the isolation of any faults in the service. In addition, it would be possible to build a coherent systemwide performance data set that could prove extremely valuable to researchers attempting to model network performance and new experimental Diffserv services. Each QBone domain was to provide a web site for disseminating and presenting its measurements. Both [MRTG]-style summary plots and raw measurement data were to be made available through this HTTP interface. The QBone architecture specified a simple, uniform URL name space; canonical names for all measurements nodes and metrics; and standard reporting formats for each. From this base, it would be straightforward to create rich tools for auditing, visualizing, or analyzing QBone measurement data.

Deployment plans

Initial deployment of QoS services implemented host DS field marking, with no signaling, and with some flow-recognition near the edge. To allow QBone deployment and experimentation to begin as soon as possible, reservations would initially be long-lived and would be established manually, relying on human operators to make admission control decisions, provision appropriately, and configure edge devices. This manual method of reservation adhered to a set of common operational practices agreed upon by QBone participants. However, it was expected that the complexities of the manual resource allocation, device configuration, and policy management would soon overwhelm the capabilities of a human operator.

In parallel with the initial QBone Diffserv deployment, participants in the QBone Bandwidth Broker (BB) working group then were working to define and implement bandwidth brokers, and to agree on a single inter-BB communication mechanism. It was envisioned that the introduction of BBs into the various QBone Diffserv domains would be performed in phases, based on the type of admission control functionality provided by the BBs. However, this did not actually happen, for reasons to be explained below.

Summary of plan

The QBone was the first wide-area test of the evolving differentiated services architecture, and the first experimental deployment of an inter-domain Diffserv signaling protocol. It was envisioned that the QBone would grow in scale and robustness as router vendors increasingly incorporated Diffserv functionality in their products and as the engineering of new services to support advanced applications became better understood. By building a highly instrumented testbed that was open and accessible to engineers, researchers, and advanced applications developers, QBone would advance the state of Diffserv technology and support the emerging field of Diffserv research. Finally, by working together with the broader Internet2 community to come to terms with the profound administrative, economic, and policy implications of QoS, QBone would prototype a working inter-domain deployment of IP QoS that could be transferred to the commercial Internet.

Actual Deployment Experience

Although QPS was successfully demonstrated at SuperComputing 2000, it enjoyed no operational deployment because of a variety of technical and non-technical obstacles summarized below in the "Lessons Learned" section. Nevertheless, a number of demanding advanced applications were deployed widely in Internet2, including a variety of high-end video conferencing and collaboration tools. These applications succeeded due to amply provisioned Best-Effort capacity and improvements in application design, especially congestion adaptation and packet loss concealment.

In 2001, Internet2's QoS efforts shifted to the QBone Scavenger Service ([QBSS]), a service designed to let users and applications take advantage of otherwise unused network without substantially affecting performance of the default Best-Effort class of service. QBSS is basically a lower than Best-Effort service. With QBSS, hosts mark their packets voluntarily for degradation in the presence of congestion. Network operators forward QBSS-marked packets without rewriting the DSCP and, where congestion might occur, configure a low-priority queue for QBSS traffic.

At first, QBSS appears ridiculous. However, there are a number of good reasons why users might mark their traffic voluntarily for worse service:

- A user may compete with himself for a remote network resource. For example, long-lived bulk transfers might occur over the same network path as real-time videoconferences.

- One might already self-police today (for example, by trying to run intensive applications "at night," which is to say during periods of low network use).

- Networks that impose usage-based charges would naturally charge less (possibly nothing) for QBSS.

For network operators, deploying QBSS is far simpler than deploying QPS. No policers are needed and differentiated treatment for QPSS is optional as well. Low-priority queues may be configured incrementally on just those interfaces where congestion might occur. The only firm requirement on a network is that it forward packets marked for QPSS without modifying their Diffserv code point.

QBSS was deployed in Abilene in 2001 and in several campus networks. Several bulk transport applications, including WU-FTPD and BitTorrent, were modified to support QBSS, and it was confirmed experimentally that QBSS loads had negligible impact on BE performance. However, with so much capacity and so little congestion in Internet2, there was little demand for any form of CoS, even something as simple as QBSS.

Lessons Learned

This section reads a bit like a list of grievances against the QPS service. Each "problem" discussed below can be overcome. The important question is: "at what cost?" Costs should be construed to include direct financial costs, business disruptions, and opportunity costs (for example, a less scalable/flexible future Internet architecture). Many of the problems itemized below could be overcome by incurring up-front costs, the sum of which can be thought of as a kind of "activation energy" that would be required to realize an inter-domain Premium service. Overcoming other problems, however, would require recurring costs. Still other problems, would require both up-front and recurring costs to overcome. At the highest level, Premium has failed to deploy because the cost is too high relative to the perceived benefits.

Poor incremental deployment properties

To support Premium service, a network must provide Expedited Forwarding (EF) [RFC3246] treatment for Premium traffic on all of its interfaces. Because EF must be implemented by a priority queue (or something morally equivalent to one), the network must be prepared to "crisp its edge"—to police on *all* ingress interfaces—to avoid a catastrophic EF DoS attack. Consequently, it is impossible to deploy Premium incrementally only when and where there is congestion; it must necessarily be deployed at the granularity of an entire network.

Missing Diffserv functionalities

Today's high-speed routers usually have some QoS functionality, but it is often insufficient for implementing Premium service. Simple DSCP-based traffic classification, leaky-bucket policing, and priority queueing are not sufficient. Below we describe some of the additional Diffserv router functionalities, which are required to implement Premium.

The first piece of missing functionality from our perspective is support for route-based classification. Premium-enabled network service providers will want to classify and police incoming EF traffic based on routing aggregates. To see this, observe that "firehose" policing (a single EF leaky bucket per ingress interface) results in hopelessly inefficient network use, because a provider must assume that the EF traffic from all interfaces could, in the worst case, converge on a single interior or egress interface. Also observe that micro-flow policing (one EF leaky bucket per micro-flow reservation traversing an ingress interface) unravels most of Diffserv's aggregation properties at inter-domain boundaries and would not scale in the core. Our conclusion is that Premium-enabled network service providers would want to sell "virtual trunks" between a pair of ingress and egress interfaces. Such a virtual trunk must be policed at ingress on the basis of an egress-dependent profile. For example, one would like to be able to configure an interface at a router like this:

rate-limit DSCP 46 traffic with next-hop AS of A to X bps

rate-limit DSCP 46 traffic with next-hop AS of B to Y bps

Without such hooks, maintenance and operation of Premium become very hard. We have inter-domain routing algorithms for very good reasons. Not having them for the purposes of policing and shaping is not much better for Premium service than having to do static routing would be for Best-Effort service.

Unfortunately, no high-speed router today provides route-based traffic classification. The reason is simple: Forwarding, as done by line cards, doesn't require full routing information. To reduce the price of line cards, forwarding tables provide a highly localized view of routing that usually only contains next-interface data. Caching the AS path (or a portion of it) in the forwarding tables would make routers significantly more expensive, while going to the route-processor for the AS path would make routers significantly slower.

The second piece of missing functionality is support for shallow token buckets. Premium aggregates must be smoothed to be nearly jitter-free as they traverse inter-domain boundaries. Policing such an aggregate effectively requires a classical token bucket policer that is only one or two packets deep. Few routers support token bucket policers this shallow at high line rates due to the fine-grained timing required.

The third piece of missing functionality is the capability to shape multiple aggregates within a Priority Queue (PQ) class. Because the downstream interface across an inter-domain boundary may be policing multiple EF aggregates, an egress interface must be able to accurately shape several aggregates *within* PQ class. That is to say, on the egress line card, shape several aggregates and then give them EF treatment across the link. Too often shapers are matched one-to-one with forwarding classes (for example, there is only one PQ class, and it can be shaped or not, as a whole, not individually).

Dramatic operational and economic paradigm shifts for operators

Because deployment of Premium is an all-or-nothing proposition, it requires fairly sudden and significant changes to network operations and peering agreements. On the operations side, operators must configure a lot of router features they usually ignore, must respond to admissions requests, and must provision carefully to honor the service assurances of admitted requests. Migrations to new routers or circuits must be performed with the utmost care. Finally, very rapid IGP convergence becomes essential and admissions decisions must be made with careful attention to routing (or be made so conservatively as to allow routing to be ignored).

Peering arrangements would also experience a dramatic paradigm shift. Today, a NSP's technical interface to the outside world is unicast IPv4, BGP, and possibly a simple Service Level Agreement (SLA), while its economic interface is some combination of per-line and per-bit charges. Premium service would complicate this with a series of additional external interfaces including: shaping, policing, reservation signaling, and per-reservation billing and settlement. Premium not only changes the interface between a NSP and its neighbors, but also adds whole new complexities for customer support personnel, creates the need for accurate third-party service auditing, and greatly increases the risk of litigation.

Threat to Best Effort

Why would a network provide high-quality Best-Effort service to the transit traffic of non-customers in a Premium-enabled world? To answer this question, consider why transit traffic is treated well today: (1) It is technically hard for a provider to differentiate between traffic from direct customers and traffic from its peers; (2) providing poor quality Best-Effort service for transit traffic today can help conserve resources, but would not translate into immediate monetary revenues. Nor would it improve the long-term reputation of a provider.

If QoS mechanisms become available in the routers to allow classification on the basis of AS path, reason (1) goes away. Further, in a Premium service world, making a customer that otherwise doesn't pay you directly switch from "free" (for transit customers) Best-Effort service to paid Premium service and have some money dribble through the payment system into your coffers seems too obvious a trick not to play; therefore, reason (2) would become increasingly irrelevant. We expect that if Premium were deployed, providers would begin to treat the Best-Effort traffic of non-customers worse than the Best-Effort traffic of their customers.

The erosion of Best-Effort service would lead to a completely different world where all serious work gets done over Premium service and users are generally expected to make virtual circuit reservations for most of what they do. By deploying Premium service, do we want to supplement the Internet Best-Effort service or to replace it?

Lack of flexible business model

Although Premium is specified with some flexibility about what "low loss" and "bounded delay" really mean, there has been inadequate thinking about how statistics can be brought to bear on either the engineered service assurance or the provisioning techniques [QoSrisk]. In practice, the service that is advertised and sold to the customer ("Premium service with zero loss and jitter") cannot be identical to the service that is actually engineered by the provider. Businesses built around service assurances (for example, FedEx, business class air travel, frame relay) do not strive for 100 percent service reliability. By separating the advertised service from the engineered service, these businesses have the flexibility to trade off statistical overbooking and operational corner-cutting against the probability that customer assurances will not be met and that money will have to be refunded.

To maximize profit, Premium must ultimately be explained to the customer in simple terms, but engineered carefully by the provider with a strong understanding of the statistical nature of traffic and of the network's performance. Of course, the statistical nature of traffic is always changing as new applications emerge and older ones fade away; so, this effort would have to be ongoing. There is insufficient theoretical understanding of how to do this kind of traffic modeling well for IP networks.

As discussed above, Premium service is not really about the network performance that *is* experienced by a reservation holder, but is rather about the performance that *would be* experienced by the reservation holder in the event of a network DoS attack [QoS-DoS]. That is, it's about the assurance. Consequently, an observation of zero loss and jitter on a Premium reservation over an extended period of time does not confirm that the Premium assurance is functioning correctly. Savvy customers will demand accurate assurance outage reports or strong recourse for service failures (penalties to the providers that are more severe than "your money back"). This is generally not something that service providers have accepted in other businesses; physical package delivery companies and PSTN service providers usually limit the remedies available to their customers to the amount paid.

Inadequate standardization and architectural gaps

A factor contributing to the reluctance of NSPs to deploy Premium has certainly been the confusion in the IETF Differentiated Services Working Group over several key areas of standardization. Chief among these is the EF per-hop behavior itself. The original EF PHB RFC [RFC2598] published in June 1999 by Van Jacobson et al. was shown to be unable to be implemented. Over more than a year, debate raged on within the working group about how to fix it. The result was the formation of a design team to author a new EF specification [RFC3248]. This specification was ultimately rejected in favor of a competing alternative [RFC3246], which was not published as a standards track RFC until March 2002.

A second factor was the decision that DSCP values would have local significance only. We regard this as a colossal mistake, burdening all edge routers with the need to re-mark traffic and creating a frivolous (but nevertheless confusing) choice for engineers. Although the choice not to have DSCPs with global significance hurts Premium deployment, it hurts services with nice incremental deployment properties even more [QBSS][ABE].

Finally, although some architectural gaps and ambiguities remain in the Premium design, we believe that these gaps do not constitute a leading reason for Premium's failure to deploy. Other "holes" include: the provisioning and matching of policers and shapers across inter-domain boundaries to support micro-flow aggregation; the calculation of worst-case jitter bounds; and the need for scalable, automated signaling.

Conclusion

In the United States today, the price of network capacity is low and falling (with the notable exception of residential and rural access) and the apparent one-time and recurring costs of Premium are high and rising (with interface speeds). In most bandwidth markets important to network-based research, it is cheaper to buy more capacity and to provide everybody with excellent service than it is to mess with CoS.

In those few places where network upgrades are not practical, Diffserv deployment is usually even harder (due to the high cost of Diffserv-capable routers and lack of clueful network engineers). In the very few cases where the demand, money, and networking expertise are present, but the bandwidth is lacking, ad hoc approaches that prioritize one or two important applications over a single congested link often work well. In this climate, the case for a global, inter-domain Premium service is dubious.

Internet applications are designed to degrade gracefully. TCP is a perfect example of this; audio and video codec with Error Correction Code is another. The upside is that if something in the network is not working quite correctly, the user either does not notice or does not care. The downside is that users often don't notice failures until they are catastrophic.

In a world of Guaranteed Services, applications will either rely on the guarantees provided by the network or they will continue to include code to adapt. In the latter case, non-catastrophic failures of Premium service would remain hidden and accumulate. In the former case, adaptation would atrophy and applications would lose their ability to work over "normal" Best-Effort networks. But, if this were to happen, adaptive applications would once again have a competitive advantage, offering users comparable functionality without the need to purchase Premium reservations. A world where Premium and Best-Effort services co-exist would seem to be unstable.

Finally, one has to ask: "Even if there were high demand and a compelling and stable business case for Premium, is this what we want the Internet to become?" The Internet's Best-Effort service model and the end-to-end principle [Saltzer] have allowed the Internet to become the fast, dumb, cheap, and global infrastructure that we all know and love. One has to wonder whether such a fundamental change to the Internet architecture, while enabling an undeniably useful and probably lucrative service, would conceal a huge opportunity cost: an increase in complexity that would inhibit the scalability and growth of the Internet in years to come.

The Internet2 community's success with the approach of enabling advanced applications by providing ample bandwidth to end users, and its experience with QPS, suggests that rather than introduce additional complexity and additional costs to implement prioritizing techniques, commercial Internet providers should focus on supplying an abundance of bandwidth to end users.

About the Contributors

Ben Teitelbaum is Product Manager and Product Marketing Manager at BitTorrent, Inc. Previously, Ben was at Internet2 and Advanced Network and Services, where he led various national-scale Internet technology initiatives, including: the QBone interdomain IP Quality of Service (QoS) testbed and the SIP.edu and ISN (Freenum) Voice over IP (VoIP) peering initiatives. A frequent speaker and author on emerging Internet technologies at academic and industry conferences, Ben holds degrees in mathematics from MIT and in computer science from the University of Wisconsin–Madison. Ben lives in Oakland, California, with his wife, children, and collection of antique typewriters.

Stanislav Shalunov is an Internet protocols expert and the author of the One-Way Active Measurement Protocol standard (RFC4656). Stanislav was a Senior Internet Engineer at Internet2 for seven years, chairing the Bulk Transport working group, contributing to standardization through the IETF and to the development of the Internet2 network, particularly its QoS aspects, leading several software development projects, and providing technical expertise for policy discussions on net neutrality in the U.S. Senate and elsewhere. He currently lives in San Francisco and works as Director of Engineering at BitTorrent, creating the next-generation transport protocol for peer-to-peer networks. Stanislav's degree in mathematics is from Moscow State University.

CASE STUDY 2: DELIVERING QOS AT INTERNAP (BY RICKY DUMAN)

Ricky Duman
Internap Network Services Corp.
rduman@internap.com

Introduction to Internap

Internap Network Services Corporation (http://www.internap.com/) is an Internet solutions company that provides route-optimized IP networking and Quality of Service (QoS) for applications and rich media content. The company provides application hosting, IP networking, and Content Delivery Network (CDN) services

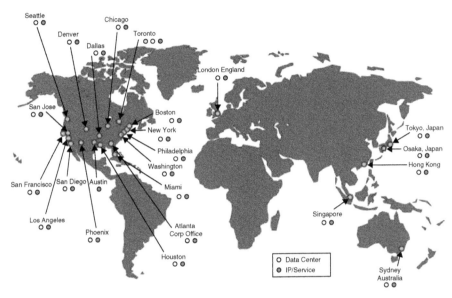

FIGURE 12-2

Internap network architecture

through more than 40 data centers and Private Network Access Points (P-NAPs) in North America, Europe, and Asia. Core to its delivery capabilities is a patented technology called Managed Internet Route Optimizer (MIRO), which monitors global Internet backbones and automatically selects the best route based on performance and reliability. The global infrastructure of data centers and Points of Presence (PoPs) are interconnected by other carriers' networks and monitored using Internap's management systems. The high-level network architecture is diagrammed in Figure 12-2 below.

Internap's customers are primarily companies that require high performance for their Internet-based applications. Internap has more than 3500 enterprise customers including significant representation in the financial services, healthcare, technology, government, retail, travel, and media/entertainment markets. Conceptually, Internap can be considered as a thin layer over the Internet. It picks the best performing part of the Internet to deliver its customers' traffic.

Internet Performance Problems

The Internet has been a tremendous success. But it also has a number of performance challenges. These challenges are described below.

Problem 1: Lack of Performance Consideration in Interdomain Routing

The Internet consists of many network domains. Each domain is called an Autonomous System (AS). The ASs are glued together by a protocol called Border Gateway Protocol (BGP). BGP helps each AS to understand which destinations are at which AS, and through what sequence of ASs to go through to get to each destination.

To work well for a large system like the Internet, BGP was designed with scalability as the primary concern. This largely mandates simplicity. Consequently, BGP uses a simple criterion to select the route for each destination—the route with the fewest ASs is considered the best. This assumes that all ASs are equal from a performance perspective.

But not all ASs are equal in performance. Some ASs cover a much larger geographic area and involve more internal links and nodes. Therefore, the propagation delay, queueing delay, jitter, and PLR can all be greater. At certain times, a particular AS may experience some performance issues, for example, internal congestion, resulting in larger latency, jitter, and packet loss ratio than multiple ASs combined. Therefore, the shortest AS path does not necessarily have the best performance. When multiple AS paths have the same length, the path picked by BGP may not have the best performance either, because BGP has no mechanism to account for performance. In other words, the BGP designers basically traded off performance consideration for scalability consideration.

For some companies, the Internet with the performance-agnostic BGP works fine. But for companies that depend on the Internet for their web sites, applications, and content, the BGP approach can be inadequate.

Problem 2: Unpredictable Performance for Distant Users

Today the Internet covers the whole world. When the content users are far away from the content providers, packets carrying the content have to travel over a long distance. This results in a large propagation delay. In addition, long distance generally means that many network hops will be involved. Every hop can introduce some latency, jitter, and packet loss. Therefore, the performance for distance users can be unpredictable. For companies that care about their users' quality experience, for example, Internet video providers, it is desirable to have this limitation removed or relieved.

Internap's Performance Solutions

Internap's performance solutions are designed to address the two issues described above. The first issue is addressed by the BGP Route Optimization solution. The second issue is addressed by the Content Delivery Network (CDN) solution.

Route optimization

In essence, route optimization enables interdomain routing to take performance into consideration in route selection. Because BGP provides a number of control knobs to allow operators to specify preferences to control the path selection, such knobs can be used to make BGP select the desired routes. This produces the desired effect on the network.

In order to take performance into consideration in route selection, three things are needed. First, performance of multiple alternative routes must be measured; second, performance-based route selection must be enabled; third, performance reporting must be available. Route optimization basically provides these three things.

Before we explain Internap's route optimization solution, we would like to briefly address why it is not desirable for network operators to do this manually.

First, a large amount of work is involved. The network operators must set up the performance measurement and performance reporting infrastructure. This requires a large amount of work.

Second, manual manipulation can be too slow. For human beings, it takes time to analyze the problem, make the decision, and execute the decision. By the time these are done, the performance problem may have already caused much damage.

Third, it is not scalable. With 200,000+ IP address prefixes in the routing table, it is virtually impossible to manually figure out which prefixes have performance problems, let alone how to address such problems.

Internap's route optimization solutions come in two forms. The premise-based solution is called Flow Control Platform (FCP). The network-based solution is called Performance IP.

Flow Control Platform (FCP)—Premise-Based Route Optimization Appliance

The FCP is an appliance. An enterprise or Internet Service Provider (ISP) that is already multi-homed to multiple carriers can connect it to their edge router to enable route optimization. The usage scenario is depicted in Figure 12-3.

The FCP solution consists of three subsystems:

- FlowCollector
- FlowDirector
- FlowView

They are described below.

FlowCollector For optimum redundancy, the FCP appliance sits out of the path of traffic, connected to a span port off of the edge switch infrastructure, as shown in Figure 12-3. The span port sends a copy of all the traffic transiting

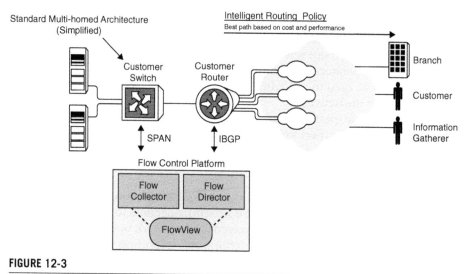

FIGURE 12-3

Flow Control Platform

the edge switches to the FCP so that it can monitor every flow's performance. FCP establishes predefined thresholds for performance variances for each flow. Should a flow to an end destination exceed a latency or packet loss threshold, the FCP FlowDirector module will put that destination into an active probing state.

FlowDirector The FlowDirector processor has a probe budget of 10,000 simultaneous probes, which could include a number of different probe types, such as:

- VIP Routes—Designated by the user of the FCP
- Sick Routes—Routes exceeding performance thresholds
- Top Volume Routes—By traffic volume
- New Routes—Newly originated flows

Performance statistics, for example, latency, jitter, and packet loss ratio, are collected for these routes.

The FlowDirector processor also peers with the edge routers via Internal BGP (iBGP). Based on the network administrator's predefined FCP policies, the FlowDirector processor could instruct the edge routers to move traffic to a different route, based on performance, cost, deep packet inspection, or capacity thresholds metrics. In the unlikely event of hardware failure or power outage, the FCP route changes would be withdrawn, and the edge routers would converge to a prespecified BGP configuration. The FCP solution can also be deployed in pairs to maximize availability.

FlowView The FlowView tool is the visibility and reporting engine, providing summary-level and detailed reports on network performance and cost. It is accessed by a management port on the appliance, which can be configured for internal and external network access. Access lists are supported on the interface to prevent unwarranted access. The FlowView tool provides over 100 intelligent reports, giving carriers visibility into key Internet metrics, including:

- Bandwidth utilization and performance by carrier
- Route change summary by carrier
- AS-level packet volumes, latency, and packet loss
- Prefix-level packet volumes, latency, and packet loss
- Historical performance data to maintain a trouble history
- Cost reports based on actual carrier billing plans
- Detailed route change activity

An example FlowView report (Last Hour Summary) is illustrated in Figure 12-4. Within this report you can see that 285 route changes were recommended in order to improve performance characteristics. This is a classic example of why manual route optimization is not feasible, and why this practice is best left to an automated approach.

Performance IP—Network-Based Route Optimization

Although FCP is fairly simple to use for a network professional, it does require connectivity to multiple ISPs and can be a distraction from the core IT operations of the company. For these companies, Internap provides the Performance IP managed network service using the Private Network Access Point (P-NAP) infrastructure and Managed Internet Route Optimization (MIRO) technology. A P-NAP is a cluster of routers and switches multi-homed to six to eight tier-1 providers. As a result, each P-NAP has six to eight possible routes to every destination on the global Internet. The customers simply connect their router to Internap's P-NAP, and Internap will perform the route optimization for them. The usage scenario is shown in Figure 12-5.

Similar to how the FCP works, P-NAP uses its MIRO route optimization technology to inject performance metrics into the route selection process. MIRO accomplishes this by automatically discovering where customers are routing to and dynamically establishing a probing list of 250,000 destinations in the top 50,000 prefixes. MIRO then monitors latency and packet loss characteristics at the route level, and performs route changes in near real time. This enables Internap's customers to always use the best performing routes for their traffic.

Route-optimized Content Delivery Network

Internap's route-optimized Content Delivery Network (CDN) is a solution for the Internet performance problem caused by distance. Internap puts content servers in its data centers and PoPs worldwide. CDN allows content to be replicated

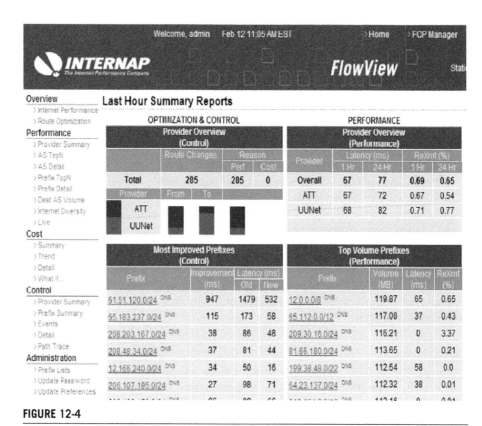

FIGURE 12-4

FlowView report: Last Hour Summary

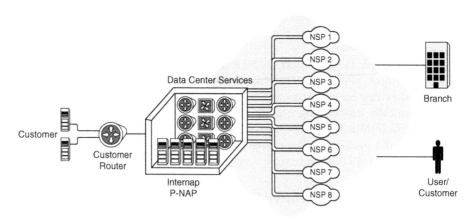

FIGURE 12-5

Internap Performance IP Service

across data centers, and served to the end users from the closest data center. This effectively moves the content closer to the users, thus reducing the performance issues caused by long distance. By using CDN, content providers can get global QoS without having to build out a global infrastructure.

Internap's CDN is different from other service providers' CDNs in that it is route optimized. This strikes a good balance between performance and cost. With traditional CDNs, the providers either have to build many data centers across the world to get the content close to the end users, or have to sacrifice performance somewhat to control the number of data centers to control cost. Therefore, their customers either have to pay more for the large number of data centers and replications, or have to live with whatever performance the limited data centers can provide. With Internap's route optimized CDN, the dilemma is solved.

Results

With Internap's P-NAP infrastructure, MIRO technology, and CDN solutions, the company offers some of the most aggressive Service Level Agreements (SLAs) in the industry. For Performance IP, the metrics are:

Availability

- 100 percent, excluding local access within North America
- 100 percent from North America to Europe
- 99.7 percent from North America to Japan
- 99.7 percent from Europe to Japan

Packet Loss Ratio

- <0.3 percent on average within North America
- <1 percent on average from North America to Europe
- <1 percent on average from North America to Japan
- <1 percent on average from Europe to Japan

Latency

- <45 ms round trip on average within North America
- <115 ms RT on average from North America to Europe
- <150 ms RT on average from North America to Japan
- < 325 ms RT on average from Europe to Japan

Jitter

- <0.5 ms on average within North America
- <2 ms on average from North America to Europe
- <5 ms on average from North America to Japan

In comparison, the typical SLAs in the industry will have

- Availability 100 percent
- Packet loss ratio, 0.5 percent
- Latency, 50–70 ms
- Jitter, 1–2 ms

For CDN services, the SLA provides 100 percent uptime and performance metrics with third-party validation by Keynote Systems.

In summary, Internap provides unmatched latency, jitter, and packet loss ratio performance in the industry. More importantly, other carriers' claim of 100 percent network availability is really just an objective. Every network can suffer an outage which will affect their customer's services. When that happens, what a carrier does is provide a refund to the affected customers, typically one day's service fee for each hour of outage. For customers who do business on the Internet, their revenue and reputation loss from such outage can far exceed the refund. In contrast, by connecting to multiple tier-1 carriers, Internap can switch traffic away from the carrier that suffers the outage. In addition, the P-NAP itself has a high degree of redundancy. The protocols for the devices inside the P-NAP are simpler and fewer in number than those in a WAN network. For example, there is no need for MPLS or Diffserv inside a P-NAP. Given that network failures are mostly caused by these complex protocols in WAN, and Internap has multiple WAN routes from which to choose. Internap provides far superior network availability.

As for the FCP, because it is an appliance sold to a customer, it does not change the SLA between the customer and its service providers. However, by choosing the Internet routes intelligently for the customer, FCP can typically reduce network latency by an average of 35 percent, and reduce bandwidth costs by 20 to 50 percent (by choosing the lower cost service providers as much as possible).

Conclusion

When discussing Internet QoS, many people automatically think of Diffserv. However, the commercial success of Diffserv QoS is rare. In contrast, through Internap's route optimization and CDN services, enterprises are able to access unmatched network availability, latency, jitter, and packet loss ratio performance. To the best of our knowledge, this is one of the most successful commercial QoS offerings on the Internet.

About the Contributor

Ricky Duman is a graduate of the University of Florida with degrees in Electrical Engineering and Computer Engineering. He has held the position of Technical Consultant for Internap Network Services for four years. Prior to that, he held multiple network engineering positions at Sprint. Ricky resides in South Florida with his wife, and enjoys fishing in the Keys.

SUMMARY

The key points of this chapter are:

- Internet2 attempted to make use of Diffserv EF PHB to provide a Premium service that would make an IP connection appear like a wire. During the process they discovered many practical issues. Internet2 eventually concluded that it is more economic to rely primarily on capacity to provide good service quality, possibly with some localized traffic prioritization in the place where capacity is not available;
- By making use of route control and CDN, Internap manages to provide one of the most aggressive SLAs in the industry in terms of network availability, delay, delay variation, and packet loss ratio. Internap prices QoS into the Premium Internet Service that they sell, and has succeeded in generating revenue from QoS.

The Internet2 case study showed that using Diffserv and traffic management to provide QoS can have many practical challenges. The Internap case study showed that providing Qos does not necessarily need Diffserv or traffic prioritization.

Please voice your opinion about this chapter at http://groups.google.com/group/qos-challenges.

QoS in Wireless Networks

13

Dr. Vishal Sharma
Principal Consultant & Technologist, Metanoia, Inc., v.sharma@ieee.org

Dr. Abhay Karandikar
Professor, Dept. of EE, IIT Bombay, karandi@ee.iitb.ac.in

In this chapter, we will examine QoS in wireless data networks, that is, Wi-Fi and WiMAX networks. The purpose is to give readers a feeling of what wireless QoS is like, thus further broadening the QoS big picture.

Because wireless networks have very different characteristics from wireline networks, QoS mechanisms and practices in wireless networks are also significantly different from those in wireline networks. Readers who are not interested in learning wireless QoS can jump to the conclusion chapter directly.

The theme of this chapter may also appear somewhat different from the rest of the book. The terms QoS and CoS are not as distinctive as in other chapters. This is partly because wireless networks have very different characteristics, and partly because it is contributed by Dr. Sharma and Dr. Karandikar, not by the author. The purpose of this chapter is to identify key underlying technical mechanism for providing QoS in wireless networks so that readers can appreciate how wireless QoS is different from wireline QoS. Again, diversified perspectives are considered a good thing. The readers can get a more balanced view before forming their own opinion.

This chapter is organized as follows. First, we start by discussing the differences between wireless and wireline networks, especially regarding QoS. Second, we examine the key QoS features and techniques for 802.11 Wi-Fi networks. Third, we examine the key QoS features and techniques for 802.16 WiMAX networks. To keep our discussion focused on key issues, we limit our exposition to 802.16d static WiMAX networks, and hint at some of the additional issues that arise in 802.16e mobile WiMAX networks. Fourth, we compare and contrast the distributed wireless QoS model exemplified by IEEE 802.11 and the centralized QoS model exemplified by IEEE 802.16. At the end, we offer some concluding remarks.

HOW QoS DIFFERS IN WIRELESS AND WIRELINE NETWORKS

Wireless networks have fundamental differences relative to wireline networks, making the operation of QoS mechanisms in wireless networks more complex than in wireline networks.

Wireless networks' fundamental differences from wireline networks include:

- Wireless channels are less reliable, because of multipath fading, interference, etc. This leads to more effort in optimizing physical transmission capability.

- Network resource is more limited, e.g., because of limited spectrum. This leads to more emphasis on allocating network resource carefully, exploiting the nature of the wireless channel.

- Mobile devices have less processing power and higher requirement on energy efficiency. This leads to the use of only processing and energy efficient mechanisms.

These points are elaborated below.

Transmission Constraint

In a wireless environment, the received signal strength attenuates with increasing distance between the transmitter and receiver. Also, obstructions, by large objects such as buildings or hills, cause the signal strength to decay in a random fashion. Moreover, reflections and scattering from various objects typically cause several copies of the transmitted signal to reach the receiver via multiple paths. This leads to large variations in received signal strength over small time scales. In a multipath fading environment, the relative motion between the transmitter and receiver may also cause broadening of the power spectrum. Additionally, the wireless channel, or the over-the-air link between a transmitter and a receiver, is subject to interference from other users in the system.

The net effect of these variations of the wireless channel is a lower signal-to-noise ratio, which leads to a high error rate and a reduction in the effective data rate. The typical bit error rate in wireless networks may be of the order of 10^{-7} compared to a wireline network, where it may be as low as 10^{-12}. Moreover, errors in wireless networks are bursty in nature, which are more difficult to correct than random bit errors. In wireline networks, the amount of data lost because of corruption during transmission is negligible, and may be handled by retransmission. By contrast, wireless networks require special strategies to combat fading at the physical layer, and intelligent scheduling and retransmission schemes to enable resource allocation at the link and network layers.

Spectrum Constraint

Wireless spectrum is scarce. The amount of bandwidth available for wireless users is generally one or multiple orders of magnitude lower. Due to limited wireless spectrum, each user must share the available spectrum in an efficient manner. Thus, the design of a spectrally efficient signal and multiple access technique is very important for wireless QoS.

Energy Constraint

Wireless devices, particularly handheld and mobile devices that are battery operated, are energy constrained because they lack a continuous source of power. This means that the algorithms and protocols for wireless transmission must be energy efficient. The devices may additionally be constrained by their processing capability, making the design of such systems even more challenging.

Therefore, the goal of wireless QoS is to optimize use of limited resources and meet the requirements of different applications by providing ways to control access to, and use of, the wireless medium, based on the characteristics of the application. Since each application has its own requirements for delay, bandwidth, jitter, and loss, the QoS capability must cater to these needs. For example, applications that require high reliability, such as data or email that need delivery of error-free files, must be delivered with low loss and error rates. Similarly, applications that require low delay (for example, voice) may be given higher priority to use the medium, while applications that require higher bandwidth (for example, video) may be assigned longer transmit times or more efficient modulation schemes at the physical layer.

In the next two sections, we examine how these issues are tackled in two popular wireless standards, the IEEE 802.11 Wi-Fi standard for wireless LANs, and the IEEE 802.16 standard for wireless MANs.

QOS IN WI-FI NETWORKS

The popular Wireless Fidelity (Wi-Fi) LAN standard (more accurately, the IEEE [802.11] standard) is based on the Carrier Sense Multiple Access with Collision Avoidance (CSMA/CA) technique. At its heart, this standard defines a short-range access technology, with a typical access point having a range of under 100 meters (about 300 feet) and the ability to handle on the order of a dozen simultaneous connections. The original standard specifies both the Physical (PHY) layer as well as the Media Access Control (MAC) layer. Wi-Fi technology enables the deployment of wireless access in a cheap and flexible way, and, over the last decade or so, has found widespread use in campuses, hotels, airports, offices, hospitals, stock markets, factory floors, and many other enterprise environments.

The first version of the [802.11] standard was released by the IEEE in 1997, with three PHY options, all of which supported data rates of 1 to 2 Mb/s. In 1999, the

IEEE released two higher-rate PHY extensions to the original standard: 802.11b, which operates in the 2.4 GHz band and provides data rates of up to 11 Mb/s, and 802.11a, which uses Orthogonal Frequency Division Multiplexing (OFDM) technology, operates in the 5 GHz band, and provides data rates of up to 54 Mb/s. Finally, in 2003, the IEEE released the 802.11g standard, which operates in the 2.4 GHz band and provides data rates of up to 54 Mb/s. With the explosive growth of Wi-Fi in corporate settings, and its need to support complex applications, requiring increasing amounts of bandwidth leads to work on IEEE's 802.11n standard, which was designed to increase by a factor of 4 the theoretical maximum speed of 54 Mb/s currently possible with the 802.11g standard. More information about these standards can be found at [802.11abgn].

The IEEE 802.11n standard, which was begun back in the 2004 timeframe, is currently very close to completion (expected to be ratified in November 2008, and formally released in June 2009), and promises impressive performance gains over 802.11g. It does so by introducing some new features and tweaking existing ones to extract the maximum performance from this technology.

At the PHY layer, 802.11n uses OFDM technology (akin to 802.11g), but combines it with Multiple-Input, Multiple-Output (MIMO) concepts. MIMO uses multiple antennas (say 2 or 4) at the transmitter and receiver to transmit several data streams in parallel, thus allowing more data to be transmitted in a given period of time, and boosting data rates. Additionally, MIMO also increases the range of transmission due to the spatial diversity provided by multiple antennas. This is coupled with adaptive channel coding, which uses special coding at the transmitter/receiver to realize the full capacity of a MIMO link.

Table 13-1 provides a summary of the key IEEE 802.11 PHY standards together with their maximum expected data rates, frequency bands of operation, and the PHY layer technologies used.

With the shift to more bandwidth-intensive and performance-sensitive applications, such as VoIP, video conferencing, and multi-media delivery, over WLAN networks, the need for Quality of Service (QoS) support in Wi-Fi networks has grown significantly in the last 3–4 years. Providing such QoS support in 802.11-based networks proved difficult, however, since the original 802.11 standards (IEEE 802.11, 802.11a, 802.11b, and 802.11g) had essentially no provision for traffic differentiation or QoS, and consequently, provided less than optimal performance for voice and video applications. It was the IEEE 802.11e-2005 standard [802.11e] that introduced notions of traffic priority and queueing that enables different types of traffic to experience different service.

Wi-Fi Operation

To understand QoS in Wi-Fi networks, it is important to first briefly understand the operation of the basic 802.11 standard. The MAC layer, whose main function is to coordinate access to the wireless medium, is the main element in enabling QoS capability.

Table 13-1 Summary of various 802.11 PHY Layer standards released to date

IEEE 802.11 PHY Protocol	Release date	Max. data rate	Operating frequency band	PHY technology
802.11 (Legacy)	1997	2 Mb/s	2.4 GHz	IR, FHSS, or DSSS
802.11a	1999	54 Mb/s	5 GHz	OFDM
802.11b	1999	11 Mb/s	2.4 GHz	DSSS
802.11g	2003	54 Mb/s	2.4 GHz	OFDM
802.11n	June 2009 (expected) Draft to be finalized in Nov. 2008	248 Mb/s		OFDM with MIMO (2 Tx, 2 Rx or 4 Tx, 4 Rx), and adaptive channel coding

Since collisions are difficult to detect in a wireless environment, the Wi-Fi standard uses a backoff-based collision avoidance technique (CSMA/CA) instead of the more familiar Carrier Sense Multiple Access/Collision Detection (CSMA/CD) used in standard Ethernet. The standard specifies two channel access (or medium access coordination) mechanisms: a mandatory Distributed Coordination Function (DCF), and an optional Point Coordination Function (PCF).

Distributed Coordination Function (DCF)

The DCF uses a collision avoidance mechanism to control access to the shared wireless medium. This is because collisions are difficult to detect in a wireless environment, so a backoff-based collision avoidance technique, rather than the collision detection technique common in standard Ethernet, is used.

Each wireless station/user first listens to the wireless medium to detect transmissions. If the medium is sensed to be busy, the user waits until the ongoing transmission is over. If the medium is detected to be idle for a Distributed Inter-Frame Space (DIFS) interval, the user enters a backoff procedure. In the backoff procedure, the user selects a random backoff time (in slots) from a contention window, and starts decrementing a backoff counter for each slot that is sensed to be idle. If, while counting down, another user begins transmitting, the user in backoff mode suspends its counting, until the transmitting user finishes and the medium is sensed to be idle for a DIFS duration, and resumes its countdown thereafter. Once the backoff interval expires, the user begins transmission. The value of the random backoff interval is chosen from an interval called the Contention Window (CW), which lies between two preconfigured values, CW_min and CW_max. The contention window is set to CW_min at the first transmission attempt, and doubles after each unsuccessful attempt, until it reaches CW_max (after which it

FIGURE 13-1

Event sequence in DCF CSMA/CA

remains at CW_max). The contention window is reset to CW_min after every successful transmission. This procedure is illustrated in Figure 13-1.

After transmitting its data, the user waits for a Short-Inter-Frame Space (SIFS) for an ACK from the recipient, which notifies the user that its frame was successfully received. If the ACK is not received within a time out period, the sender assumes that there was a collision, and schedules a retransmission by entering the backoff process again, until the maximum number of retransmissions has occurred, at which point the packet is dropped.

Point Coordination Function (PCF)

The PCF is a polling-based access scheme, with no contention. In PCF, the access point takes control of the medium, acting as a point coordinator. The access point divides the channel access time into periodic beacon intervals. A beacon interval is the period of time between the sending of beacon frames by an access point, which communicate network management and identification information to all user stations in a specific WLAN. Each beacon interval comprises of a Contention Period (CP) and a Contention-Free Period (CFP). During the former, DCF, as described above, is used. During the latter, the stations wait to be polled, with the access point acting as the polling master. The access point maintains a list of active stations, and polls them in a round-robin fashion.

Since the PCF and DCF co-exist, the access point waits for a PCF Inter-Frame Space (PIFS) before starting PCF. The PIFS interval is made shorter than DIFS to give the PCF higher access priority than DCF. The complete 802.11 media access event sequence is illustrated in Figure 13-2.

The PCF was an optional MAC mechanism, because its hardware implementation was considered to be rather complex at the time when the standard was ratified, circa 1997–1999. As a result, it ended up not being widely implemented in commercial products, and is not used much today.

Wi-Fi QoS Model

We observe that neither access mechanism outlined above has any explicit QoS functionality. This is not surprising, since Wi-Fi technology evolved primarily to provide wireless access for basic data services in enterprise environments. At the time of its development and for several years thereafter, best-effort data services were sufficient to meet the needs of enterprise and residential users. However, starting a few years ago, users have begun wanting voice, video, and audio services to be delivered

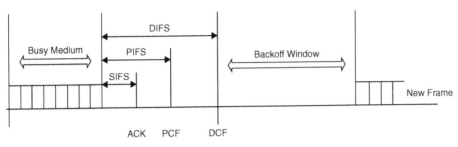

FIGURE 13-2

Event sequence in 802.11

through WLAN connections. The need to unwire also has become pervasive in both enterprises and residences. The 802.11 standards (a/b/g), however, were not designed to provide the QoS that such advanced services require, as we discuss next.

QoS limitations of 802.11a/b/g

The DCF, being contention-based, gives access to the medium to the first packets that grab it. This creates a fairness problem, and can potentially starve some stations for long periods of time. Also, it provides only a best-effort service that treats all traffic the same, with every station and traffic type having equal priority. This becomes a problem for streamed data and real-time applications, such as voice/video, as contention causes drops that affect quality.

Although PCF eliminates collisions and the time spent on backoff and/or contention, it still suffers from the drawback that it has no way to distinguish between different types of traffic. It also has no mechanism to provide priority in access to the wireless medium, nor any way for the user/end station to communicate its QoS requirements to the access point.

A problem common to both the DCF and the PCF is that the 802.11 legacy MAC does not include any admission control mechanism, so under high traffic loads the performance of both functions degrades, and all traffic suffers.

In summary, the QoS limitations of the 802.11a/b/g are:

- No support for CoS.
- No support for user/end stations to communicate their QoS requirements to the access point.
- No support for admission control.

QoS capabilities enabled by 802.11e

The IEEE 802.11e standard is an enhancement to the MAC sublayer to add QoS functionality to Wi-Fi networks. It does so by adding support for:

- Prioritizing data packets based on their type.
- Allowing user/end stations to communicate their QoS requirements to the access point.
- Supporting admission control.

These basically address the QoS limitations of the 802.11a/b/g.

In IEEE 802.11e, a new MAC layer access technique called the Hybrid Coordination Function (HCF) is introduced. It replaces the DCF access technique of 802.11a/b/g technologies. The HCF has two access methods:

- Enhanced Distributed Channel Access (EDCA), a contention-based mechanism useful for giving QoS to data traffic.

- HCF Controlled Channel Access (HCCA), a polling-based mechanism useful for giving QoS to voice and video traffic.

A key feature of the HCF is the notion of a Transmission Opportunity (TXOP), which represents the time duration during which a station is allowed to transmit a burst of data frames.

Enhanced Distributed Channel Access (EDCA)

The EDCA mechanism is the contention-based part of HCF. It enhances DCF to support CoS. EDCA creates four Access Categories (ACs) using the [802.1p] field. Each AC is mapped to a transmit queue. Akin to the IEEE 802.1p, 802.11e can support 8 priorities, which are mapped to 4 ACs, as shown in Table 13-2.

Upon entering the MAC layer, each data packet received from a higher layer is tagged with a specific priority (0 to 7). This tagging algorithm is left to individual implementations, and is not standardized.

In short, EDCA is DCF done on a per-AC basis. For each AC, four key parameters are varied to introduce differentiation among the ACs:

- Minimum contention window size, CW_min.

- Maximum contention window size, CW_max.

- Transmission opportunity limit (TXOP), which specifies the maximum length of time that an 802.11e-enabled node/user can transmit.

Table 13-2 Priority to Access Category mapping in 802.11e

Priority	AC	Traffic type
0	0	Best-Effort
1	0	Best-Effort
2	0	Best-Effort
3	1	Video Probe
4	2	Video
5	2	Video
6	3	Voice
7	3	Voice

- Arbitration Inter-Frame Spacing (AIFS), which defines the time interval between the wireless medium becoming idle and the start of channel access negotiation (akin to DIFS in DCF).

So, a set of four parameters CW_max[AC], CW_min[AC], TXOP[AC], and AIFS[AC] are defined for each AC. These parameters are similar to DCF's, except for the introduction of TXOP. Generally speaking, smaller CW_min, CW_max, and AIFS will lead to higher probability of transmission, that is, higher priority, while larger TXOP will lead to higher bandwidth. By setting these parameters accordingly, some differentiation can be created.

Simulation results show that the differentiation provided by such a categorization leads to markedly different delay and throughput for different ACs, thus allowing for differentiation between the classes [Y.Xiao][Q.Ni]. However, even with such differentiation, the performance of video and data flows can be degraded when the channel is heavily loaded. Real-world deployment of EDCA is still limited.

HCF-Controlled Channel Access (HCCA)

The HCCA mechanism is the contention-free part of HCF. Just like the EDCA is an enhancement of DCF in 802.11a/b/g, the HCCA is an enhancement of the PCF. The differences between HCCA and PCF are:

- HCCA may take place in both the contention period and the contention-free period, while PCF can only take place in the contention-free period.

- HCCA allows user/end stations to communicate their QoS requirements to the access point and supports admission control.

By using a polling-based mechanism, HCCA reduces medium-access contention. HCCA has the ability to poll stations if it detects the medium to be idle. Therefore, it cuts down the time when the medium goes unused. Further, it is able to support scheduling of data based on a station/user's traffic flow requirements (for example, desired min/peak data rate, delay, and service interval, and the user's average/min/peak packet size), which are communicated to the access point using a Traffic Specification (TSPEC) in a QoS request frame. The access point then applies an admission control mechanism to admit or deny a TSPEC request. The centrally scheduled access mechanism requires that the client knows in advance the resources that it needs and communicates that to the access point so that it may schedule concurrent traffic effectively.

To leave sufficient room for EDCA to operate, the maximum duration for which HCCA can operate in a beacon interval is bounded by a system parameter that is set during initial WLAN configuration.

Summary of Wi-Fi QoS

In summary, QoS control in 802.11 looks like Figure 13-3.

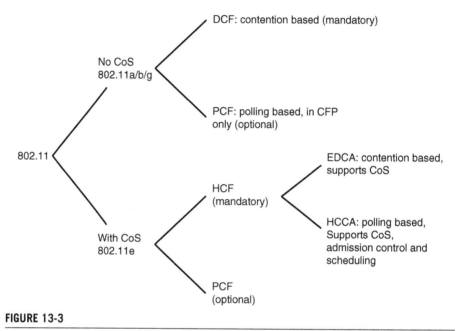

FIGURE 13-3

QoS control in 802.11

The key points for Wi-Fi QoS are:

- The original 802.11 family of standards (802.11a/b/g) included no explicit QoS support, only a contention-based DCF mode and a polling-based PCF mode.

- In the contention-based DCF mode of operation, the only QoS mechanism was collision avoidance. The lack of traffic classification mechanisms, the contention-based channel access, and a simple FIFO packet scheduler implied that there was no way to distinguish between different types of traffic and no service guarantees possible. In the polling-based PCF mode, contention was largely eliminated, but the problem of traffic differentiation remained.

- For an 802.11 end user this effectively translated to only best-effort service. While this was not an issue for data applications, such as email and file transfer, it could render applications such as VoIP and video ineffective at moderate to heavy network loads, for example, at hotspots or in heavy-use corporate environments.

- Driven by these limitations and the increasing performance expectations of users of Wi-Fi networks (as they began using complex applications and services), the 802.11e standard attempted to retro-fit the legacy 802.11a/b/g Wi-Fi standards with QoS capabilities. This was achieved via the new HCF mechanism, which included a contention-based EDCA suitable for data applications, and a polling-based HCCA suitable for voice/video applications.

- The major enhancement provided by EDCA was the introduction of eight traffic priorities and four access categories (ACs), and priority packet scheduling, which gave preference to the four ACs in decreasing order of importance. The ability to dynamically separate packets from a given source into different ACs by classifying them using the 802.1p tag or DSCP value. The parameters: AIFS, CW min/max, and TXOP enabled the four ACs to differentiate themselves, providing significant performance gains, especially for high-priority traffic.

- The major enhancement provided by HCCA was the ability to provide QoS to voice/video applications by accounting for the QoS parameters when allocating resources to different stations that were being polled. To an end user both of the above implied that it was now possible to differentiate between different traffic types—voice, video, data (ftp, email), and even provide quantifiable QoS to voice/video applications (via HCCA).

- In both of the above, the PCF remained an optional capability, which was not widely embraced in implementations, and so ended up not being widely used in deployments either.

Besides the EDCA and HCCA, the IEEE 802.11e standard provides several other MAC enhancements to more efficiently use the wireless medium. The first is the introduction of contention-free bursts, which allows a station/user to send a series of frames in a row, without having to contend for the channel again. As long as there is time remaining in a TXOP, the station only waits for a SIFS delay, and continues transmission. This reduces overheads of DIFS and backoffs. Since the original 802.11 standard requires each unicast data frame to be immediately acknowledged, the 802.11e standard introduces the notion of a block ACK, which increases efficiency by aggregating ACKs for multiple frames into a single ACK. Additionally, the new standard allows for data to be piggybacked on polling requests and ACKs, thus improving network efficiency.

A few points regarding Wi-Fi QoS are worth mentioning. The original need for QoS in Wi-Fi networks stemmed from the use of VoIP and multi-media applications in corporate environments. To solve this problem, simple prioritization of packets using proprietary means had begun in the mid-to-late 2002 time frame, long before the 802.11e standard was fully developed and ratified. The prioritization was also available on a number of handset uses for VoIP access over WLANs. This enabled some form of QoS in both the downlink (access point to end station) and uplink directions. For the most part, this simple prioritization worked, since at low to moderate network loads, it gave higher priority to voice traffic, and maintained acceptable quality.

The sophisticated QoS mechanisms, such as those in HCF, were an attempt to provide "carrier-grade" service in the Wi-Fi environment, and, in a practical sense, remain futuristic. Indeed, a number of products today use variants of their proprietary solutions, coupled with software to monitor and configure access points, which can provide a QoS capability as good as (and, in some cases, even better than) what is possible by the enhancements outlined in the standards.

QoS IN WIMAX NETWORKS

While Wi-Fi networks have been a popular choice in homes, offices, and hotspots like hotels and airports, these networks are not designed for high-speed mobile and outdoor access. The IEEE has defined the 802.16 family of standards [802.16] for broadband wireless access. The Worldwide Interoperability for Microwave Access (WiMAX) Forum has been formed primarily to promote the adoption of products based on the 802.16 standard. WiMAX is proving to be a good alternative to other last-mile broadband access solutions such as DSL, cable, and fiber. The high cost of installation and maintenance of the cable plant makes WiMAX an attractive option, particularly in the emerging markets and greenfield deployments where the cable/copper infrastructure is either poor or does not exist.

The WiMAX standard defines a high-bandwidth, long-range technology, with aggregate bandwidth as high as 40–70 Mbps per channel (shared among multiple users) and a cell radius in the neighborhood of 10 km. The large coverage area provided by WiMAX makes it an appealing choice for providing broadband connectivity in sparse rural areas as well.

The IEEE 802.16 WG was formed in 1998. Initially, it was chartered to develop the standard for broadband access in the 10–66 GHz range, which supports line-of-sight operation. The target application was enterprise-class services, such as T1/E1 leased-line replacements, video conferencing, and packetized voice. The physical layer was based on single carrier modulation.

Subsequently, the scope of the standard was enhanced to include residential-class deployment with non-line-of-sight operation in the 2–11 GHz range, which added Orthogonal Frequency Division Multiplexing (OFDM) as the physical layer. The final standard was published as IEEE 802.16–2004. This standard also incorporated support for Orthogonal Frequency Division Multiple Access (OFDMA) as one of the multiple access modes.

It soon became apparent to the designers of the standard that mobility support would be an important component for the standard to compete favorably with other mobile technologies. Thus, support for mobility was added, in the form of IEEE 802.16e-2005 [802.16]. The standard also supports multiple antennas (i.e., MIMO) to improve the spectral efficiency of the system. Currently, the IEEE 802.16m WG is considering support for high-speed mobility (200 km/hour and beyond) to further enhance the capabilities of the 802.16e standard. Thus, 802.16 is a collection of standards with various modes of operations. The standard supports selectable channel bandwidths from 1.25 MHz to 20 MHz. Though WiMAX supports 2-11 GHz for non-line-of-sight applications, the 2.5 GHz, 3.5 GHz and 5 GHz frequency bands have found most favor for a majority of practical deployments.

An important aspect of the WiMAX standard is that support for QoS has been an integral part of the design of its MAC and PHY layers right from the beginning. QoS was not an afterthought as with Wi-Fi. Many of the QoS features of WiMAX were adopted from well-developed ATM and DOCSIS frameworks, with enhancements made for wireless communications.

WiMAX Operation

To understand the role of QoS in WiMAX networks, it is useful to first understand the basic operation of WiMAX networks. A WiMAX network consists of a fixed Base Station (BS) and multiple subscriber stations. In WiMAX, time is divided into frames. The length of a frame can vary from 0.5 msec to 20 msec, with 1 msec and 5 msec being the most common. Each frame is divided into a downlink (from BS to subscriber station) subframe and an uplink subframe. Each subframe is further divided into a fixed number of slots. A subscriber station transmits data in specific slots designated by the BS. Conceptually, a slot represents a unit of transmission bandwidth. Transmission and QoS are achieved by allocating the slots to various subscriber stations accordingly, taking the need of the applications into consideration (for example, voice needs low delay/jitter, FTP needs high bandwidth).

What is a slot?

The slot is determined by the multiple access technique such as Time Division Multiple Access (TDMA) or Orthogonal Frequency Division Multiple Access (OFDMA). In a TDMA system, a slot is a segment of time allocated to a given user for transmitting its data. In an OFDMA system, a slot is basically one or multiple unique (time slot, frequency) combinations.

How are slots allocated?

In order for a subscriber station to get some slots from the BS, it must first send the bandwidth requests. In WiMAX, bandwidth requests are normally transmitted in one of the two modes: a contention mode and a contention-free mode (polling).

In the contention mode, the subscriber stations send bandwidth requests during a contention period. The BS assembles all requests and computes a conflict-free schedule taking into account the resource requirements, the wireless link quality to each subscriber, etc. Grants from the BS are communicated to subscriber stations in control slots in the downlink subframe. Because multiple subscriber stations can send bandwidth requests simultaneously, contention can happen. It is resolved using an exponential backoff strategy. That is, after sending their requests, those stations that do not receive a grant back off in a binary-exponential way before contending again.

In the contention-free mode, the BS polls each subscriber station and a subscriber station in reply sends its bandwidth request. The polling interval must be such that the delay requirements of the various classes of traffic/services can be met.

WiMAX QoS model

The designers of WiMAX have adopted a QoS model similar to ATM's. That is, the system uses a connection-oriented architecture. This is different from that in Wi-Fi or Ethernet, both of which follow a connection-less model. At application setup, the subscriber establishes a connection with the BS. The advantage of a connection-oriented architecture is that the resource allocator can associate a particular connection with a set of QoS parameters and traffic profiles. The classification of IP/Ethernet/ATM packets to 802.16 packets has been defined in the 802.16 standard. While the QoS model is similar to ATM's, because of the dynamic nature of the wireless link, the following differences exist:

1. Dynamic resource allocation: In WiMAX networks, resource allocation can be done on a frame-by-frame basis. In other words, a connection can get different amount of resource at different times.

2. Adaptive Modulation and Coding (AMC): WiMAX allows for link quality feedback to be available to the transmitter, so that it may select appropriate burst profiles for transmission. A burst profile is a combination of modulation, coding, and forward-error correction schemes that is used for data transmission. For example, if the link quality is very poor, the transmitter can fall back to more robust modulation and coding schemes. This will cause the data to be sent at a lower rate, but will ensure that it is received correctly. This minimizes the impact of errors on the wireless link. The available modulation methods in 802.16d and their peak transmission rates are described in Table 13-3.

Apart from the above fundamental characteristics, several other key QoS supporting features were built into the system, as explained below.

First is the capability to have flexible framing. This is needed to optimally use the available airtime on the wireless link by allowing a subscriber or base station to adapt to changing conditions on the wireless link. For example, the relative portion of a frame devoted to uplink/downlink transmission can be dynamically changed from frame to frame depending on the traffic need. The relative portion of control/data in sub-frames can also be dynamically changed depending on the network load. This can make the use of the wireless spectrum more efficient.

Second is the ability to provide symmetric high-throughput in both uplink and downlink directions. This is achieved by having scheduling intelligence at both the subscriber and the base station, and by allowing the subscriber stations to

Table 13-3 Modulation and coding schemes for IEEE 802.16d

Modulation	Overall coding rate	Peak data rate in 5 MHz (Mbps)
BPSK	1/2	1.89
QPSK	1/2	3.95
QPSK	3/4	6.00
16-QAM	1/2	8.06
16-QAM	3/4	12.18
64-QAM	2/3	16.30
64-QAM	3/4	18.36

communicate their QoS requirements to the base station. The symmetric throughput is important because WiMAX networks are likely candidates for numerous backhaul applications that require the transport of aggregated traffic in both directions.

Third is the capability for frame packing and fragmentation. These are effectively QoS techniques because they allow WiMAX systems to pack many small-sized packets into a frame, or to break up a large packet into multiple frames. This helps to prevent a high-priority packet from having to wait a long time for the completion of transmission of a large low-priority packet. This also enables a more effective use of the varying bandwidth of the wireless link. For example, when the link between the user and the BS is in good condition, a large volume of data (in the form of packed frames) can be rapidly exchanged. When the link is in poor condition, packets may be fragmented and transmitted to meet some minimum bandwidth goals, giving the wireless medium for other users whose links to the BS are in good condition.

In summary, the WiMAX standard provides the following key QoS features:

1. Dynamic resource allocation to adapt to the varying condition of the wireless link and real-time need of the application.
2. Adaptive modulation and coding to minimize wireless link errors and increase throughput.
3. Flexible PHY/MAC framing to maximize the bandwidth available for actual data transmission.
4. Support for symmetrical high throughput in both uplink and downlink directions
5. Efficient transport of MAC frames via packing and fragmentation.

QoS services

With these key QoS features, the IEEE 802.16 standard defines four services with different performance requirements. Each service is associated with a set of performance parameters. The four types of services are:

- Unsolicited Grant Service (UGS)
- Real-time Polling Service (rt-PS)
- Non-real-time Polling Service (nrt-PS)
- Best-Effort service (BE)

These services are mandatory, in the sense that any higher-layer application will have to be mapped to a WiMAX connection that belongs to one of the above four services. Thus, any standards-compliant WiMAX system must implement the above services.

It is worth noting that WiMAX does not confine an implementation to having only four classes. Nor does it impose any specific queueing implementation at the subscriber station or the base station. An operator/vendor may have multiple classes of traffic, and queue them in any manner, e.g., per connection, per application, or per class.

Unsolicited Grant Service (UGS)

This service supports real-time data streams consisting of fixed-size data packets transmitted at periodic intervals, such as voice-over-IP without silence suppression. It is analogous to ATM's CBR service. The mandatory QoS service-flow parameters for this service are:

- Maximum Sustained Traffic Rate,
- Maximum Latency,
- Tolerated Jitter, and
- Request/Transmission Policy.

These applications require a fixed-size grant on a real-time periodic basis. Thus, to a subscriber station with an active UGS connection, the BS provides a fixed-size data grant at periodic intervals based on the Maximum Sustained Traffic Rate of the service flow. The Request/Transmission Policy for UGS service is set such that the subscriber station is prohibited from using any contention request opportunities for a UGS connection. To reduce bandwidth consumption, the BS does not poll the subscriber stations with active UGS connections. If a subscriber station with an active UGS connection needs to request bandwidth for a non-UGS connection, it may do so by setting a bit in the MAC header of an existing UGS connection to indicate to the BS that it wishes to be polled. Once the BS detects this request, the process of individual polling is used to satisfy the request.

Real-time Polling Service (rt-PS)

This service supports real-time data streams consisting of variable-sized data packets that are transmitted at fixed intervals, such as MPEG video. It is analogous to ATM's rt-VBR service. The mandatory QoS service-flow parameters for this service are:

- Minimum Reserved Traffic Rate
- Maximum Sustained Traffic Rate
- Maximum Latency
- Request/Transmission Policy

For this type of service, the BS must provide periodic unicast request opportunities that meet the real-time needs of the flow. In these request opportunities, the subscriber station can specify the size of the desired grant. The request overhead for this service is more than that of UGS, but it supports a variable grant size, thus improving its transmission efficiency. The Request/Transmission Policy is set such that the subscriber station is prohibited from using any contention request opportunities for such connections.

Non-real-time Polling Service (nrt-PS)

This service supports delay-tolerant data streams, consisting of variable-sized data packets for which a minimum data rate is required, such as web browsing. It is analogous to ATM's nrt-VBR service. The mandatory service flow parameters for this type of service are:

- Minimum Reserved Traffic Rate
- Maximum Sustained Traffic Rate
- Traffic Priority
- Request/Transmission Policy

Such applications require a minimum bandwidth allocation. The BS must provide unicast request opportunities on a regular basis that ensure that the service receives request opportunities even during network congestion. The Request/Transmission Policy is set such that subscriber stations are allowed to use contention request opportunities. Thus, the subscriber station can use contention request opportunities as well as unicast request opportunities to request bandwidth.

Best-Effort (BE) service

This service is designed to provide efficient transport for best-effort traffic with no explicit QoS guarantees. The mandatory service flow parameters for this service are

- Maximum Sustained Traffic Rate
- Traffic Priority
- Request/Transmission Policy

Table 13-4 802.16 services classes and their characterizing parameters

Service	Applications	Parameters
UGS	Uncompressed voice, TDM circuits	■ Maximum Sustained Traffic Rate ■ Maximum Latency ■ Tolerated Jitter ■ Request/Transmission Policy
rt-PS	Video, VoIP	■ Minimum Reserved Traffic Rate ■ Maximum Sustained Traffic Rate ■ Maximum Latency ■ Request/Transmission Policy
nrt-PS	Web browsing, interactive data applications	■ Minimum Reserved Traffic Rate ■ Maximum Sustained Traffic Rate ■ Traffic Priority ■ Request/Transmission Policy
Best Effort	Email, FTP	■ Maximum Sustained Traffic Rate ■ Traffic Priority ■ Request/Transmission Policy

These applications do not require any minimum service level and, therefore, can be handled on a "space available" basis. These applications share the remaining bandwidth after allocation to the previous three services has been completed. The Request/Transmission Policy is set such that the subscriber station is allowed to use contention request opportunities.

In summary, the four service classes, their target applications, and the performance parameters are described in Table 13-4.

A few points regarding WiMAX QoS are worth noting:

First, the traffic scheduler at the BS decides the allocation of the physical slots to each subscriber station on a frame-by-frame basis. While making allocations, the scheduler must account for the following:

- Scheduling service specified for the connection
- Values assigned to the connection's QoS parameters
- Queue sizes at the subscriber stations
- Total bandwidth available for all the connections

Second, although requests to the BS are made on a per-flow basis, the grants by the BS are issued on a per-subscriber station basis, without distinguishing individual flows at the subscriber station. This means that there needs to be a local scheduler at each subscriber station that allocates the bandwidth grants from the BS among the competing flows. This model was adopted in 802.16 for two reasons. (1) Granting on a per-subscriber station basis reduces the amount of state information that the BS must maintain. (2) Since the local and link conditions can change dynamically, having per-subscriber station grants allows a subscriber station scheduler flexibility to assign resource to more important new flows.

Third, the scheduler used at the BS/subscriber station, a critical component of the WiMAX QoS architecture, is not specified in the standard. It is up to the implementers to decide. Therefore, this is an area of considerable research [Iyengar] [Rath1][Rath2][Sharma1][Sharma2].

Summary of WiMAX QoS

The various design features incorporated in the IEEE 802.16 PHY and MAC layers, as well as the careful design of the scheduling services that mirror ATM's, allow for parameterized QoS to be given to individual flows in a WiMAX network.

However, we note that although the standard has attempted to build sophisticated QoS capabilities into WiMAX, it does not imply that WiMAX will necessarily provide better QoS than other wireless technologies. While WiMAX certainly has the mechanisms to do so, and current research and analysis show positive effect, the end result will depend on how these capabilities are embraced by service providers. Furthermore, there are aspects of QoS (for example, scheduling) that are not specified by the standards, for which service providers will have to make choices. A lot will depend on the choices made and their efficacy.

As practical experience with WiMAX deployments grows, the industry will have a better understanding regarding how to effectively use WiMAX's sophisticated QoS features.

WI-FI QOS VERSUS WIMAX QOS

Wi-Fi and WiMAX networks have adopted fundamentally different approaches to providing QoS. This stems from their roots.

Wi-Fi is a LAN technology, originally designed to provide wireless access for office/residence environments. With the passage of time, however, enterprise wireless networks began to be used for a lot more than just email or data. VoIP, video conferencing/streaming, etc. over Wi-Fi LANs are perceived to create the need for QoS capabilities. This was why IEEE 802.11e was started.

WiMAX, on the other hand, is a MAN technology designed for providing carrier grade services. As such, making provisions for QoS was a fundamental design requirement from the beginning. It was envisioned that WiMAX would need the ability to provide per-flow QoS. This required several key design features, as outlined previously.

In Wi-Fi, the centralized mechanisms PCF and HCCA exist but are not widely implemented. Therefore, the Wi-Fi QoS model practically adopts the distributed approaches DCF and EDCA, where each access point and subscriber station acts independently. The distributed nature of the QoS model makes it easy to deploy, since little or no coordination between access points and stations is required. At the same time, it necessitates the use of CSMA/CA, which is subject to collisions and can impose significant access overhead and waste precious bandwidth, as much as 30 percent at high loads.

Since it was not originally designed with QoS capabilities in mind, the QoS feature set available in Wi-Fi is limited to prioritization. This allows for relative performance differentiation among different types of traffic. Guaranteeing absolute measures of performance is impossible.

WiMAX, by contrast, has a very rich QoS feature set that is modeled after ATM's, and provides for parameterized QoS. Thus, one can specify precisely the bandwidth, delay, jitter, and loss requirements for individual flows, and the schedulers at the BS/subscriber station will work to honor those requirements at the wireless link. In WiMAX, the scheduling and multiple-access are centralized at the BS, which means the BS can efficiently schedule transmissions without wasting bandwidth, unlike Wi-Fi. Such scheduling also implies that the access overhead for a node to access the medium is practically negligible, since each subscriber station knows exactly when it should do so, and does not have to rely on trials and collisions.

However, the strong QoS capabilities in WiMAX come at a price, higher complexity for implementation and deployment. One may be tempted to draw an analogy between WiMAX QoS and ATM QoS, and between Wi-Fi QoS and Ethernet QoS. One may even be tempted to conclude that WiMAX QoS may suffer the same fate of ATM QoS. However, wireless and wireline networks do have significant differences. Above all, network resource is much more limited in wireless networks. When network resource is limited, it is easier to justify sophisticated resource allocation methods such as WiMAX's. Time will tell whether WiMAX's strong QoS effort will pay off.

SUMMARY

The key points of this chapter are

- Wireless networks pose significant challenges in resource allocation relative to wireline networks due to constraints arising from: the use of a shared medium, interference, lower channel reliability, and limited energy.

- Resource allocation in wireless networks must consider the characteristics of the physical layer.

- Wi-Fi networks primarily employ a contention-based access mechanism, where access coordination is distributed. QoS enhancement in Wi-Fi networks is, therefore, focused on traffic prioritization.

- WiMAX networks employ a centralized access coordination mechanism. A WiMAX base station can exploit link-quality information from each user to make intelligent decisions to schedule a user and to select the user's transmission parameters, such as modulation and coding types. WiMAX also provides flexible PHY/MAC framing to optimize resource allocation. WiMAX employs a connection-oriented QoS model, which is similar to ATM's. The WiMAX QoS model is more sophisticated than the Wi-Fi QoS model, but it also has higher complexity. Time will tell how well the WiMAX QoS model will fare in practice.

ABOUT THE CONTRIBUTORS

Dr. Vishal Sharma is an international technology consultant, telecom expert, researcher, and speaker with over sixteen years of experience in networking and telecom technologies, spanning network and systems design, architecture, analysis, and optimization. He has worked on the development of Quality of Service algorithms/protocols for both wireline (IP, Ethernet, ATM) and wireless (WiMAX, 3G) networks, is an expert in the analysis and design of switch/router scheduling and flow-management algorithms, MPLS technologies, traffic engineering principles, protocols, carrier approaches, planning tools, and algorithms, and in next-generation telecommunication network design—encompassing, among others, key issues such as inter-provider QoS, carrier Ethernet, and WiMAX-based broadband. He has served clients on four continents ranging from tier 1–3 providers and large/small equipment vendors, to premier technology and software houses and chip manufacturers.

Dr. Sharma is a Senior Member of the IEEE, a Fellow of the IETE, a Subject Matter Expert at the MFA Forum, on the Scientific Committee of the prestigious MPLS World Congress, and WiMax Congress. He has over 120+ talks and publications (conference and journal papers, industry standards, industry presentations, invited talks) to his credit, and 5 U.S. patents granted, with over 6 pending.

He obtained his B.Tech. degree from the Indian Institute of Technology Kanpur (IITK, 1991), and his M.S. (Signals & Systems, 1993), M.S. (Computer Engineering, 1993), and Ph.D. (Electrical and Computer Engineering, 1997) degrees from the University of California at Santa Barbara.

Dr. Sharma has been actively working on WiMAX technology for the last several years, having co-developed scheduling algorithms and techniques for providing QoS in WiMAX networks (published in numerous international conferences: Globecom, Milcom, ICC), and having delivered multiple tutorials and workshops on the subject to a broad audience in the United State and abroad over the last couple of years (e.g., Globecom'06, Milcom'07). He continues to research, and consult in, this emerging technology, as part of Metanoia, Inc.'s focus on broadband access technologies.

Dr. Sharma has lived and worked in both the United States and India, and has had professional engagements in ten countries overall.

Dr. Abhay Karandikar obtained his B.E. in 1986 and M.Tech and Ph.D. from IIT Kanpur, India, in 1988 and 1995, respectively. Between 1988 and 1989, he worked in Indian Space Research Organization, Ahmedabad. He was Member Technical Staff and Team Coordinator in High Performance Computing Group of Centre for Development of Advanced Computing (C-DAC), Pune, India, an Indian government initiative in Supercomputer during 1994 to 1997.

Since 1997, he has been working in the department of electrical engineering of IIT Bombay as a faculty member. In IIT Bombay, he led many technological development projects including an open-source MPLS router. He also co-founded the start-up company Eisodus Networks. Dr. Karandikar has supervised many graduate theses and published several papers in national and international journals and conferences. He has consulted extensively for industries in the area of communications network and wireless communications. He has served on technical program committees of many international conferences. He has received numerous awards including the Best Teacher award of IIT Bombay. His research group is currently focused on cross-layer resource allocation in wireless networks and Quality of Service guarantees in the Internet.

Conclusion

14

QoS has multiple important aspects: technical, commercial, and regulatory. Without a big picture, QoS can be like the old metaphor of the elephant to the blind—what it is like depends on which part you touch. Because modern people are highly specialized, few people have the opportunity to develop a comprehensive view on QoS. As a result, contemporary QoS wisdom has been too focused on CoS and traffic management, and has not paid sufficient attention to some other important factors, especially the commercial and regulatory ones.

On the commercial side, contemporary QoS wisdom takes it for granted that if NSPs offer a higher CoS than Best Effort, some business or consumers will see the higher CoS's value and will buy it. Several important questions have not been seriously investigated. For example, "Can customers really see the value of a higher CoS?," "What kind of assurance should NSPs provide to attract business/consumers to buy QoS without incurring too much risk for themselves?," "Will the attempt to sell QoS trigger customer defection, because it is considered as evidence of poor quality for the baseline service?" The lack of investigation on these questions reflects a lack of commercial consideration for QoS.

On the regulatory side, contemporary QoS wisdom takes it for granted that if NSPs can provide a higher CoS, they can sell it without regulatory trouble. The possibility of government intervention has not been considered until the Net Neutrality debate started.

Even on the technical side, contemporary QoS wisdom overlooks a number of key factors. First, it assumes that traffic congestion will exist somewhere, and therefore traffic prioritization is necessary for providing QoS. Whether traffic congestion can be avoided by proper network planning, or prevented/alleviated by other traffic control schemes such as traffic engineering, has not been carefully investigated. This caused the traditional QoS solution to overemphasize Diffserv and traffic management, and underemphasize other traffic control schemes. Second, the impact of device failures and operator errors on QoS has not been sufficiently taken into account. Consequently, traditional QoS solution overemphasizes the value of various traffic control mechanisms, and overlooks the

247

harm caused by their complexity. Third and most fundamentally, contemporary QoS wisdom assumes that if two classes of traffic are put into different queues and applied different traffic management treatments, the differentiation between them will be perceivable by the end users. There is little effort to verify this fundamental assumption. But, of course, if the differentiation is not clearly perceivable by the end users, Diffserv/CoS will not matter much.

Because of these problems, QoS is far from becoming a reality, from a commercial perspective.

This book strove to provide a QoS big picture. It covered all three important aspects of QoS: technical, commercial, and regulatory. The key points are:

- QoS is most meaningful when it is defined from end users' perspective. That is, QoS is good network performance for end users' applications.
- With this definition, QoS is naturally more than CoS and traffic management. Any schemes that enhance the end users' quality of experience can all be QoS tools. These schemes include CDN, interdomain and intradomain traffic engineering, network planning and auditing, traffic protection and restoration, etc., in addition to traffic management.
- When all schemes are considered, it becomes clear that it is difficult to differentiate a higher CoS to a point where the end users can clearly perceive the difference from Best Effort. Because IP traffic grows very fast, IP networks are generally provisioned with a certain amount of idle capacity to accommodate the rapid growth. Consequently, in normal network condition, Best Effort service in developed countries can already provide good performance for most applications. In abnormal network conditions caused by failure, various control schemes can take advantage of the idle capacity to redistribute traffic to avoid or alleviate congestion. This can allow Best Effort to continue to maintain good enough performance. Real world Internet performance measurement results confirmed this. Network failures may cause QoS issue, but CoS cannot address the failures themselves. Lack of interprovider QoS settlement further compounds the difficulty to differentiate.
- Lack of perceivable differentiation between a higher CoS and Best Effort leads to difficulty to sell the higher CoS. Users will not be able to see its value. This causes all the commercial challenges to sell QoS explicitly. If NSPs push ahead to sell it, users will either defect to other NSPs, or demand government intervention if there is no other NSP to go to. The latter will cause the regulatory challenges like Net Neutrality.

Knowing that the differentiation difficulty is the fundamental cause of other technical, commercial, and regulatory challenges, this book concluded that QoS should be priced into other services (e.g., IPTV) and be sold implicitly as part of those services. CoS can still be used inside the network to optimize performance, but should be transparent to the end users. Because QoS is embedded in other services and is not sold separately, the commercial challenges associated with selling it largely disappear. Because there is no traffic discrimination based on whether a QoS fee is paid, the regulatory challenges also disappear. This model also simplifies the technical QoS solution, because NSPs can rely on capacity, network planning/auditing, and necessary controls to provide good quality for all services without worrying about creating user perceivable differentiation.

The revised QoS business model represents a simpler pricing scheme, compared to the traditional model. History showed that when a communication service is sufficiently cheap and frequently used, simple pricing schemes generally out perform complex ones. As Internet service has become part of people's daily life, we believe that the revised model is better suited for the network industry. The revised technical solution will also lead to simpler network operations and simpler network equipment.

Because QoS has been discussed for so many years, many people have hardened their QoS opinion, despite the lack of a QoS big picture. Some people simply take some fundamental issues for granted or assume that somebody else will take care of these fundamental issues. However, when the fundamental assumptions are not carefully verified, the whole theory built over them can be like a building over sand. By putting all the fundamental issues on the table and examining whether and how they are being addressed, this book hopefully makes the reality of QoS clearer. We sincerely hope that the big picture content and the revised business model provide some new perspectives on QoS.

List of Acronyms

Abbreviation	Meaning	Abbreviation	Meaning
ABR	Available Bit Rate	BS	Base Station
AC	Access Category	CA	Collision Avoidance
ACK	Acknowledgement	CAC	Call Admission Control
AF	Assured Forwarding	CAPEX	Capital Expenditure
AIFS	Arbitration Inter-Frame Spacing	CBR	Constant Bit Rate
AMC	Adaptive Modulation and Coding	CD	Collision Detection
ARQ	Automatic Repeat reQuest	CDN	Content Delivery Network
AS	Autonomous Systems	CFP	Contention Free Period
AS_PATH	Autonomous System Path	CIR	Committed Information Rate
ATM	Asynchronous Transfer Mode	CLI	Command Line Interface
BB	Bandwidth Broker	CLP	Cell Loss Priority
BE	Best Effort	CoS	Classes of Service
BECN	Backward Explicit Congestion Notification	CP	Contention Period
BER	Bit Error Rate	CSMA/CA	Collision-Sense Multiple Access with Collision Avoidance
BFD	Bi-directional Forwarding Detection	CSMA/CD	Carrier Sense Multiple Access with Collision Detection
BGP	Border Gateway Protocol	CW	Collision Window
BPSK	Binary Phase Shift Keying	DAMA	Demand Assigned Multiple Access
BRAS	Broadband Remote Access Server	DCF	Distributed Coordination Function

Abbreviation	Meaning	Abbreviation	Meaning
DDOS	Distributed Denial Of Service	HCCA	HCF Controlled Channel Access
DE	Discard Eligibility	HCF	Hybrid Coordination Function
Diffserv	Differentiated Service	HDTV	High Definition TV
DIFS	Distribute Inter-Frame Spacing	H-QoS	Hierarchical QoS
DL	Downlink	HTTP	Hypertext Transfer Protocol
DRR	Deficit Round Robin	ICP	Internet Content Provider
DSCP	Differentiated Service Code Point	IEEE	Institute of Electrical and Electronic Engineers
DSL	Digital Subscriber Line	IETF	Internet Engineering Task Force
DSLAM	DSL Access Multiplexer	IGP	Interior Gateway Protocol
DUIC	Downlink Usage Interval Code	ILEC	Incumbent Local Exchange Carrier
EDCA	Enhanced Distributed Channel Access	IMS	IP Multimedia Subsystem
EF	Expedited Forwarding	IntServ	Integrated Services
ETSI	European Telecommunications Standards Institute	IP	Internet Protocol
FCC	Federal Communications Commission	IPDV	IP Packet Delay Variation
FCH	Frame Control Header	IPG	Inter-Packet Gap
FCP	Flow Control Platform	IPTD	IP Packet Transfer Delay
FEC	Forward Error Correction	IPTV	Internet Protocol television
FECN	Forward Explicit Congestion Notification	IS-IS	Intermediate System-to-Intermediate System
FR	Frame Relay	ISP	Internet Service Provider
FTP	File Transfer Protocol	ITU	International Telecommunication Union
GDP	Gross Domestic Product	LAN	Local Area Network
GSM	Global System for Mobile communications (originally from Groupe Spécial Mobile)	LLU	Local Loop Unbundling
H-ARQ	Hybrid Automatic Repeat reQuest	LSP	Label Switched Path

Abbreviation	Meaning	Abbreviation	Meaning
MAC	Media Access Control	PDU	Packet Data Unit
MAN	Metropolitan Area Network	PHB	Per-Hop Behavior
MED	Multi-Exit Discriminator	PHY	Physical Layer
MIMO	Multiple-Input, Multiple-Output	PIFS	PCF Inter-Frame Spacing
MIRO	Managed Internet Route Optimizer	PLC	Packet Loss Concealment
MOS	Mean Opinion Score	PLR	Packet Loss Ratio
MPEG	Motion Picture Experts Group	P-NAP	Private Network Access Point
MPLS	Multi-Protocol Label Switching	PNNI	Private Network-Network Interface
NANOG	North America Network Operator's Group	PON	Passive Optical Network
NCP	Network Control Program	PoP	Point of Presence
nrt-PS	Non-Real Time Polling Service	POS	Packet over SONET/SDH
nrt-VBR	Non-real-time Variable Bit Rate	PQ	Priority Queue
NSP	Network Service Provider	PSTN	Public Switched Telephone Network
NTT	Nippon Telegraph and Telephone	QAM	Quadrature Amplitude Modulation
NVP	Network Voice Protocol	QBSS	QBone Scavenger Service
Ofcom	Office of Communications	QoE	Quality of Experience
OFDM	Orthogonal Frequency Division Multiplexing	QPS	QBone Premium Service
OFDMA	Orthogonal Frequency Division Multiple Access	QPSK	Quadrature Phase Shift Keying
OPEX	Operations Expenditure	RACS	Resource and Admission Control Subsystem
OPT-E-MAN	Optical Ethernet Metro Area Network	RED	Random Early Detection
OSPF	Open Shortest Path First	RFI	Request for Information
P2P	Peer To Peer	RFPs	Requests for Proposal
PARC	Palo Alto Research Center	RLC	Radio Link Control
PCF	Point Coordination Function	RNG	Ranging

Abbreviation	Meaning	Abbreviation	Meaning
RR	Round Robin	TXOP	Transmission Opportunity
RSVP-TE	Resource Reservation Protocol with Traffic Engineering Extension	UBR	Unspecified Bit Rate
rt-PS	Real-Time Polling Service	UGS	Unsolicited Grant Service
RTT	Round Trip Time	UL	Uplink
rt-VBR	Real-time Variable Bit Rate	UNE-P	Unbundled Network Equipment-Platform
SAR	Segmentation and Reassembly	UNI	User Network Interface
SIFS	Short Inter-Frame Spacing	VC	Virtual Circuit
SLA	Service Level Agreement	VLAN	Virtual LAN
SLAC	Stanford Linear Accelerator Center	VoD	Video on Demand
SPQ	Strict Priority Queueing	VoIP	Voice over Internet Protocol
TCO	Total Cost of Ownership	VPN	Virtual Private Network
TCP	Transmission Control Protocol	WAN	Wide Area Network
TDD	Time Division Duplex	WDRR	Weighted Deficit Round Robin
TDMA	Time Division Multiple Access	WFQ	Weighted Fair Queueing
TE	Traffic Engineering	Wi-Fi	Wireless Fidelity
TISPAN	Telecoms & Internet converged Services & Protocols for Advanced Networks	WiMAX	Worldwide Interoperability for Microwave Access
TOS	Type of Service	WLAN	Wirless LAN (802.11 based)
TSPEC	Traffic Specification	WRED	Weighted Random Early Detection

Sample Peering Contract

B

Private Interconnection Agreement
(Bilateral Peering Agreement)

This agreement (the "Agreement") is made, effective as of "date 1" (the "Effective Date"), by and between NSP1, a Delaware Corporation with an office at "address 1", and _____, a Corporation with an office at _____ _____. ("Company"). Each of NSP1 and Company is a "Party" to this Agreement, and collectively they are referred to as "Parties".

WHEREAS, each of NSP1 and Company operates an Internet Network, as defined below; and

WHEREAS, the Parties wish to provide for the connection of, and exchange of traffic between, their respective Internet Networks on the terms and conditions of this Agreement.

NOW, THEREFORE, the Parties in consideration of their mutual promises, and for other good and valuable consideration, agree as follows:

1) Definitions
    a) "Internet Network" shall mean a communications network running the TCP/IP protocol and any future protocol as specified by the IETF.
    b) "Interconnection Point" shall mean any point at which the Parties agree to connect their respective Internet Networks under this Agreement. A description of all Interconnection Points is set forth on Schedule A attached hereto, which may be amended by the Parties from time to time in accordance with the "Amendment" clause below.

2) Exchange of Traffic
    a) Company agrees to exchange IP traffic over AS #____, and NSP1 agrees to exchange traffic over AS #1234.[1]

---

[1] AS #1234 is meant to be a generic AS number. This peering agreement is not from the NSP which owns AS 1234.

b) Each Party shall provide, at its own expense, a connection from such Internet Network to the Interconnection Point(s) upon a schedule to be mutually agreed. The data rate and locations at which the Parties will interconnect hereunder is set forth in Schedule A. The data rate shall not be less than that specified in the NSP1 Peering Policy document.

c) Company agrees that route prefixes announced under this Agreement will be of a maximum prefix length of 24 bits using BGP4. Aggregation of routing information will be done where possible and technically feasible, to meet this requirement and to comply with current best practices for route aggregation.

d) Except for data required for traffic analysis and control traffic, which must be examined in order for either Party to operate, control and maintain the security of their respective Internet Networks, neither Party shall monitor or capture the contents of any data or other traffic that passes through the Interconnection Points.

e) Company is:

    i) obligated to announce all of its customers' routes to NSP1 and accept all customer routes announced by NSP1;

    ii) required to implement a consistent routing policy by using the same ASN and announcing the same set of routes at all Interconnection Points using BGP4.

f) Company is not:

    i) allowed to "default" its routing decisions to the NSP1 peer router. Defaulting routing decisions to NSP1 is defined as sending traffic to NSP1 across the interconnection point which is not destined for the NSP1 network, or a customer of that network, as defined by the set of routes announced to Company by NSP1 at the Interconnection Point.

3) Term and Termination

a) This Agreement shall have an initial term of two (2) years following the Effective Date. If neither Party chooses to terminate this Agreement upon the expiration of the initial term, this Agreement shall automatically renew for a subsequent one (1) year period.

b) Either Party may terminate this Agreement upon 60 calendar days written notice to the other Party.

c) The failure to order circuits within the time periods set forth on Schedule A shall constitute a material breach.

4) Technical and Operational Matters

a) The Parties will work together during the term of this Agreement to establish mutually agreed upon performance objectives and operational procedures to enable each Party to provide the highest practical quality of service over its designated Internet Network and the interconnection provided hereunder, in a cost effective fashion. In connection therewith, the Parties shall use their reasonable efforts to minimize end-to-end packet delay.

b) Each of the Parties will use its reasonable efforts to achieve a mean time to repair of four hours or less for all outages at the Interconnection Point(s) set forth on Schedule A.

c) Each of the Parties will develop scheduled maintenance procedures that provide for notification by one Party to the other of all scheduled maintenance that could cause end-to-end connectivity loss for any user of more than fifteen minutes. Each Party agrees to give the other two calendar days advance notice for scheduled maintenance that is expected to result in 60 minutes or more of end-to-end connectivity loss.

d) Each of the Parties will use its reasonable efforts to collect during the term hereof and provide to the other Party traffic information with respect to its designated Internet Network in order to better understand the nature of the traffic through the Parties' respective designated Internet Networks. In addition, each Party shall use its reasonable efforts to track and provide the other Party with average, 95/5 and peak utilization data over the interconnection facilities set forth on Schedule A hereto.

e) There are no specific Service Level Agreements or liabilities associated with traffic carried over a Party's Internet Network, but each Party will consider in good faith network upgrades and/or redundant traffic routes on its Internet Network if requested by the other Party.

5) Payments
    a) Unless otherwise agreed in writing by the Parties, no settlement, service, or port charges of any kind for data transmission will be paid or owed by either Party to the other under this Agreement, with the exception of situations outlined in section 5b.

    b) In the event that a Peer fails to meet any of the guidelines established in the NSP1 Peering Policy after the implementation of a Private Peering relationship, NSP1 reserves the right to terminate the peering agreement by giving the appropriate notice after requesting correction by the peer, or to implement some other mutually agreeable solution. This solution may include, but is not limited to, applying port or service charges to peering interconnections in the Region(s) in which the requirements set forth in the Peering Policy are not met.

6) Customer Relations
    Each Party will be responsible for communication with its own customers with respect to its designated Internet Network. Each Party will use its reasonable efforts to notify the other promptly in writing using inter-NOC communication of all trouble reports made to it by customers of the other

Party which relate to the Internet Network of the notified Party. Each Party will independently establish the charges to its own customers for the services provided in connection with this Agreement.

7) Non-exclusivity

This Agreement is non-exclusive and shall not prohibit or restrain either Party's entry into any similar or dissimilar contract or agreement with one or more third parties.

8) Liability

a. EACH PARTY DISCLAIMS ALL REPRESENTATIONS AND WARRANTIES, EXPRESS AND IMPLIED, ARISING OUT OF OR RELATING TO INTERCONNECTION TO ITS INTERNET NETWORK OR OTHER ACTIVITIES UNDER THIS AGREEMENT, INCLUDING ANY REPRESENTATIONS OR WARRANTIES OF MERCHANTABILITY OR FITNESS FOR A PARTICULAR PURPOSE, OR PERTAINING TO THE SECURITY OR DELIVERY OF TRAFFIC, OR THAT ANY ROUTING INFORMATION OR OTHER INFORMATION PROVIDED IS ACCURATE AND COMPLETE AND DOES NOT INFRINGE ANY PATENT, COPYRIGHT, OR OTHER INTELLECTUAL PROPERTY RIGHT.

b. NEITHER PARTY SHALL BE LIABLE TO THE OTHER FOR ANY LOSS OR DAMAGE OF ANY NATURE, INCLUDING, WITHOUT LIMITATION, DIRECT, INDIRECT, SPECIAL, CONSEQUENTIAL, OR OTHER DAMAGES, LOST REVENUES, LOST PROFITS OR LOST BU.S.INESS OPPORTUNITIES ARISING FROM: (I) ANY FAILURE IN OR BREAKDOWN OF ANY FACILITIES OR SERVICES HEREUNDER, WHATSOEVER THE CAU.S.E AND HOWEVER LONG IT SHALL LAST, (II) ANY INTERRUPTION OF SERVICE, WHATSOEVER THE CAU.S.E AND HOWEVER LONG IT SHALL LAST, (III) SUCH PARTY'S SUBMITTING TRAFFIC TO OR ACCEPTING TRAFFIC FROM THE OTHER PARTY HEREUNDER, (IV) PORT, SERVICE, OR OTHER CHARGES, OR (V) ANY OTHER CIRCUMSTANCES RELATING TO THIS AGREEMENT. THIS EXCLU.S.ION OF LIABILITY APPLIES REGARDLESS OF WHETHER THE CLAIM IS BASED UPON CONTRACT, TORT, STRICT LIABILITY, BREACH OF WARRANTY OR OTHER THEORY, AND EVEN IF A PARTY HAS BEEN INFORMED IN ADVANCE OF THE POSSIBILITY OF SUCH DAMAGES, AND WHETHER OR NOT SUCH DAMAGES ARE FORESEEABLE, *PROVIDED* ALWAYS THAT DAMAGES WHICH UNDER APPLICABLE LAW MAY NOT BE EXCLUDED AND ARE NOT EXCLUDED.

c. Each Party acknowledges that this Section 8 reflects an informed, voluntary and deliberate allocation of risks (both known and unknown) arising from or related to this Agreement.

9) Regulatory Approval

   The Parties acknowledge that this Agreement, any or all of the terms hereof, may become subject to regulatory approval by various local, state or federal agencies. Should such approval be required at any time, the Parties shall cooperate, to the extent reasonable and lawful, in providing such information as is necessary to complete any required filing and to secure the required approval.

10) Assignment

   Neither Party shall transfer or assign its rights or obligation under this Agreement or transfer by way of merger, consolidation, sale of all or substantially all of its assets, without the prior written consent of the other Party, whose consent shall not be unreasonably withheld, *provided*, that any assignee shall separately confirm its agreement to its obligations hereunder.

11) Notices

   All notices between the Parties required or permitted hereunder shall be effective only if in writing and (i) hand delivered, (ii) sent by private courier, or (iii) sent by first class mail, postage prepaid, to the address specified below. Notice by facsimile, to be effective, shall be supplemented by one of the methods listed in the preceding sentence. All notices shall be effective three days after dispatch.

12) Affiliates

   Each Party is contracting on behalf of any of its Affiliates whose networks may be included among the Internet Networks interconnected under this Agreement. Each Party agrees that each such Affiliate shall be bound by all of the terms and conditions of this Agreement. For purposes of this Agreement, an "Affiliate" of a Party means an entity that controls, is controlled by, or is under common control with, that Party.

13) Entire Agreement

   This Agreement represents the entire understanding between the Parties regarding the subject matter hereof and supersedes all other prior agreements, understandings, negotiations, and discussions between the Parties with respect to this subject matter. This Agreement shall be governed by and construed in accordance with the laws of the State of New York, without regard to the conflict principles thereof.

14) Severability

   If any provision of this Agreement is held by a court of competent jurisdiction to be contrary to law, the remaining provisions of this Agreement will remain in full force and effect.

15) Amendment

This Agreement may be modified only by a written amendment signed by both Parties.

16) No Third Party Beneficiaries

Nothing contained in this Agreement shall be deemed to confer any right in any third party not a signatory to this Agreement.

17) Confidentiality

a. All information exchanged between the Parties under this Agreement or during the negotiations preceding this Agreement and relating either to the terms and conditions of this Agreement or any activities contemplated by this Agreement is confidential and neither Party shall disclose to any third Party any of the other Party's confidential information disclosed to it. Any announcement of this Agreement must be mutually agreed upon by both Parties, including the language and timing of press releases and other announcements to third Parties. No use by a Party of the other Party's name, insignia, logos, trademarks, tradenames, or service marks ("Marks") shall be made without the other Party's consent, and all such uses of the other Party's Marks shall inure only to the benefit of the Party owning the Marks.

b. The terms of the NSP1 Non-Disclosure Agreement between the Parties are hereby incorporated herein and made a part hereof.

IN WITNESS WHEREOF, the Parties have caused their respective authorized representatives to sign this Agreement on their behalf, effective as of the date written above.

[Company Name]   NSP1 Ltd..

By: _____   By: _____

Name: _____   Name: _____

Title:_____   Title: _____

Date: _____   Date: _____

U.S. Senate Bill S215, Internet Freedom Preservation Act

Internet Freedom Preservation Act (Introduced in Senate)

S 215 IS

110th CONGRESS
1st Session
S. 215

To amend the Communications Act of 1934 to ensure net neutrality.

IN THE SENATE OF THE UNITED STATES
January 9, 2007

Mr. DORGAN (for himself, Ms. SNOWE, Mr. KERRY, Mrs. BOXER, Mr. HARKIN, Mr. LEAHY, Mrs. CLINTON, Mr. OBAMA, and Mr. WYDEN) introduced the following bill; which was read twice and referred to the Committee on Commerce, Science, and Transportation

---

A BILL

To amend the Communications Act of 1934 to ensure net neutrality.

Be it enacted by the Senate and House of Representatives of the United States of America in Congress assembled,

SECTION 1. SHORT TITLE.

This Act may be cited as the 'Internet Freedom Preservation Act'.

SEC. 2. INTERNET NEUTRALITY.

Title I of the Communications Act of 1934 (47 U.S.C. 151 et seq.) is amended by adding at the end the following:

SEC. 12. INTERNET NEUTRALITY.

(a) Duty of Broadband Service Providers- With respect to any broadband service offered to the public, each broadband service provider shall—

(1) not block, interfere with, discriminate against, impair, or degrade the ability of any person to use a broadband service to access, use, send, post, receive, or offer any lawful content, application, or service made available via the Internet;

(2) not prevent or obstruct a user from attaching or using any device to the network of such broadband service provider, only if such device does not physically damage or substantially degrade the use of such network by other subscribers;

(3) provide and make available to each user information about such user's access to the Internet, and the speed, nature, and limitations of such user's broadband service;

(4) enable any content, application, or service made available via the Internet to be offered, provided, or posted on a basis that—

(A) is reasonable and nondiscriminatory, including with respect to quality of service, access, speed, and bandwidth;

(B) is at least equivalent to the access, speed, quality of service, and bandwidth that such broadband service provider offers to affili-ated content, applications, or services made available via the public Internet into the network of such broadband service provider; and

(C) does not impose a charge on the basis of the type of content, applications, or services made available via the Internet into the network of such broadband service provider;

(5) only prioritize content, applications, or services accessed by a user that is made available via the Internet within the network of such broadband service provider based on the type of content, applica-tions, or services and the level of service purchased by the user, without charge for such prioritization; and

(6) not install or utilize network features, functions, or capabilities that impede or hinder compliance with this section.

(b) Certain Management and Business-Related Practices- Nothing in this section shall be construed to prohibit a broadband service provider from engaging in any activity, provided that such activity is not incon-sistent with the requirements of subsection (a), including—

(1) protecting the security of a user's computer on the network of such broadband service provider, or managing such network in a

manner that does not distinguish based on the source or ownership of content, application, or service;

(2) offering directly to each user broadband service that does not distinguish based on the source or ownership of content, application,
or service, at different prices based on defined levels of bandwidth
or the actual quantity of data flow over a user's connection;

(3) offering consumer protection services (including parental controls
for indecency or unwanted content, software for the prevention of
unsolicited commercial electronic messages, or other similar capabilities), if each user is provided clear and accurate advance notice
of the ability of such user to refuse or disable individually provided
consumer protection capabilities;

(4) handling breaches of the terms of service offered by such broadband service provider by a subscriber, provided that such terms of
service are not inconsistent with the requirements of subsection
(a); or

(5) where otherwise required by law, to prevent any violation of
Federal or State law.

(c) Exception- Nothing in this section shall apply to any service regulated
under title VI, regardless of the physical transmission facilities used to
provide or transmit such service.

(d) Stand-Alone Broadband Service- A broadband service provider shall not
require a subscriber, as a condition on the purchase of any broadband
service offered by such broadband service provider, to purchase any
cable service, telecommunications service, or IP-enabled voice service.

(e) Implementation- Not later than 180 days after the date of enactment of
the Internet Freedom Preservation Act, the Commission shall prescribe
rules to implement this section that–

(1) permit any aggrieved person to file a complaint with the
Commission concerning any violation of this section; and

(2) establish enforcement and expedited adjudicatory review procedures consistent with the objectives of this section, including the
resolution of any complaint described in paragraph (1) not later
than 90 days after such complaint was filed, except for good cause
shown.

(f) Enforcement-

(1) IN GENERAL- The Commission shall enforce compliance with this
section under title V, except that–

(A) no forfeiture liability shall be determined under section 503(b)
against any person unless such person receives the notice
required by section 503(b)(3) or section 503(b)(4); and

(B) the provisions of section 503(b)(5) shall not apply.

(2) SPECIAL ORDERS- In addition to any other remedy provided under this Act, the Commission may issue any appropriate order, including an order directing a broadband service provider–

    (A) to pay damages to a complaining party for a violation of this section or the regulations hereunder; or

    (B) to enforce the provisions of this section.

(g) Definitions- In this section, the following definitions shall apply:

    (1) AFFILIATED- The term affiliated' includes–

        (A) a person that (directly or indirectly) owns or controls, is owned or controlled by, or is under common ownership or control with, another person; or

        (B) a person that has a contract or other arrangement with a content, applications, or service provider relating to access to or distribution of such content, applications, or service.

    (2) BROADBAND SERVICE- The term broadband service' means a 2-way transmission that–

        (A) connects to the Internet regardless of the physical transmission facilities used; and

        (B) transmits information at an average rate of at least 200 kilobits per second in at least 1 direction.

    (3) BROADBAND SERVICE PROVIDER- The term broadband service provider' means a person or entity that controls, operates, or resells and controls any facility used to provide broadband service to the public, whether provided for a fee or for free.

    (4) IP-ENABLED VOICE SERVICE- The term IP-enabled voice service' means the provision of real-time 2-way voice communications offered to the public, or such classes of users as to be effectively available to the public, transmitted through customer premises equipment using TCP/IP protocol, or a successor protocol, for a fee (whether part of a bundle of services or separately) with interconnection capability such that service can originate traffic to, and terminate traffic from, the public switched telephone network.

    (5) USER- The term user' means any residential or business subscriber who, by way of a broadband service, takes and utilizes Internet services, whether provided for a fee, in exchange for an explicit benefit, or for free.'

SEC. 3. REPORT ON DELIVERY OF CONTENT, APPLICATIONS, AND SERVICES.

Not later than 270 days after the date of enactment of this Act, and annually thereafter, the Federal Communications Commission shall transmit a report to the Committee on Commerce, Science, and Transportation of the Senate and the Committee on Energy and Commerce of the House of Representatives on the–

(1) ability of providers of content, applications, or services to transmit and send such information into and over broadband networks;

(2) ability of competing providers of transmission capability to transmit and send such information into and over broadband networks;

(3) price, terms, and conditions for transmitting and sending such information into and over broadband networks;

(4) number of entities that transmit and send information into and over broadband networks; and

(5) state of competition among those entities that transmit and send information into and over broadband networks.

References

[ABE]   P. Hurley, M. Kara, J. Le Boudec and P. Thiran, "ABE: Providing a Low-Delay Service within Best Effort", IEEE Network Magazine, Vol. 15 No. 3, May 2001.

[Akamai]   Akamai Technologies Inc., http://www.akamai.com/html/technology/index.html

[Arbor]   Arbor Networks Inc. http://www.arbornetworks.com/

[ARCNET]   Attached Resource Computer NETwork (ARCNET), http://en.wikipedia.org/wiki/ARCNET

[ARPANET]   http://en.wikipedia.org/wiki/Arpanet

[Alvarez]   S. Alvarez, "QoS for IP/MPLS Networks", Cisco Press, 2006.

[Armitage]   G. Armitage, "IP Quality of Service", Macmillan Technology Series, 2000.

[ATMtm]   ATM Forum, "Traffic Management Specification, Version 4.0", af-tm-0056.000, April 1996

[ATMpnni]   ATM Forum PNNI subworking group, "Private Network-Network Interface Spec. v1.0 (PNNI 1.0)", afpnni-0055.00, Mar. 1996.

[Availab]   I. Hussain, "Understanding High Availability of IP and MPLS Networks", Cisco Press, Jan. 2005.

[BBN]   BBN Technologies, Inc. http://www.bbn.com/

[BFD]   D. Katz, D. Ward, "Bidirectional Forwarding Detection", draft-ietf-bfd-base-06.txt, Mar. 2007

[BGPBest]   BGP Best Path Selection Algorithm, http://www.cisco.com/en/U.S./tech/tk365/technologies_tech_note09186a0080094431.shtml

[BitTorrent]   Bit Torrent, http://www.bittorrent.com/

[Bizweek]   http://www.businessweek.com/magazine/content/05_45/b3958092.htm

[Cariden]   Cariden Technologies, Inc., http://www.cariden.com/

[Caswell]   W. Caswell, "Bandwidth: How much is needed, and how much is it worth", Home Toys Article, Aug. 2002, http://www.hometoys.com/mentors/caswell/aug02/bandwidth.htm

[CBQdemo]   R. Nitzan, "Experience with a Class Based Queuing Demonstration", http://www.es.net/nesg/esnet-qbone-participation/cbq_test_paper.html, February, 1998. (The link was missing and this paper could no longer be found. Related information can be found at: http://qos.internet2.edu/wg/documents-informational/20011002-teitelbaum-qos-futures.pdf)

[Chen]   S. Chen and K. Nahrstedt, "An Overview of Quality-of-Service Routing for the Next Generation High-Speed Networks: Problems and Solutions", IEEE Network Magazine, Special Issue on Transmission and Distribution of Digital Video, 1998.

[Clark]   D. Clark, "The Design Philosophy of the DARPA Internet Protocols", ACM SIGCOMM 88, ACM CCR Vol 18, Number 4, August 1988, pages 106–114.

[Davie]   B. Davie, "Deployment Experience with Differentiated Services", ACM SIGCOMM Revisiting IP QoS Workshop: Why do we care, what have we learned? (RIPQOS), August 2003, http://caia.swin.edu.au/ripqos/bdavie-31415.pdf

[Doyle]   J. Doyle et al, "Robustness and the Internet: Theoretical Foundations", http://netlab.caltech.edu/pub/papers/RIPartII.pdf

[DSLforum]   DSL Forum, http://www.dslforum.org/

[ETSI]   European Telecommunications Standards Institute (ETSI), http://www.etsi.org/WebSite/homepage.aspx

[Evans]   J. Evans, C. Filsfils, "Deploying IP and MPLS QoS for Multiservice Networks: Theory and Practice", Morgan Kaufmann, 2007

[FCChss]   Federal Communications Commission Releases Data On High-Speed Services for Internet Access, Jul. 2005, http://www.fcc.gov/Bureaus/Common_Carrier/Reports/FCC-State_Link/IAD/hspd0705.pdf

[FCCmrc]   "FCC Identifies, Fines VoIP Blocker", Broadbandreports.com, http://www.broadbandreports.com/shownews/60996

[FedEx]   FedEx Corp., http://www.fedex.com/

[Ferguson]   P. Ferguson and G. Huston, "Quality of Service", John Wiley & Sons, 1998.

[Ferrari]   D. Ferrari, D. Verma, "A Scheme for Real-Time Channel Establishment in Wide-Area Networks", IEEE JSAC, vol. 8 no. 3, April 1990, pp. 368–379.

[Fortz1]   B. Fortz and M. Thorup, "Internet Traffic Engineering by Optimizing OSPF Weights", IEEE Infocom 2000, Mar. 2000.

[Fortz2]   B. Fortz and M. Thorup, "Optimizing OSPF/IS-IS Weights in a Changing World", www.research.att.com/~mthorup/PAPERS/papers.html

[Fraleigh]   C. Fraleigh, "Provisioning Internet Backbone Networks to Support Latency Sensitive Applications", Ph.D. thesis, Stanford University, June 2002.

[G.114]   G.114, "One-way transmission time", ITU-T, May 2003, http://www1.cs.columbia.edu/~andreaf/new/documents/other/T-REC-G.114-200305.pdf

[HTTPdif]   B. Krishnamurthy, J. Mogul and D. Kristol, "Key Differences between HTTP/1.0 and HTTP/1.1", Proceedings of the WWW-8 Conference, Toronto, May 1999.

[Huston]   G. Huston, "Internet Performance Survival Guide: QoS Strategies for Multi-Service Networks", Wiley, 2000.

[ICFA07]   L. Cottrell, ICFA SCIC Network Monitoring Report, Jan. 2007, http://www.slac.stanford.edu/xorg/icfa/icfa-net-paper-jan07/

[IETF]   Internet Engineering Task Force, http://www.ietf.org/

[IGPfast]   C. Alaettinoglu, V. Jacobson, H. Yu, "Toward Millisecond IGP Convergence", NANOG 20 presentation, Oct. 2000

[IMS]   IMS Architecture, http://www.dataconnection.com/sbc/imsarch.htm

[Internap]   Internap Network Services Corporation, http://www.internap.com/solutions/routecontrol/page1980.html

[InterQoS]   S. Amante et al, "Inter-provider Quality of Service", Quality of Service Working Group, MIT Communications Futures Program (CFP), Nov. 2006

[IPMPLSF]   IP/MPLS Forum, http://www.ipmplsforum.org/index.shtml

[IPrelia]   H. Wang, A. Gerber, A. Greenberg, J. Wang, R. Yang, "Towards Quantification of IP Network Reliability", http://www.research.att.com/~jiawang/rmodel-poster.pdf, 2006

[IPTV]   http://telephonyonline.com/mag/telecom_china_grabs_global/

[ITU]   International Telecommunication Union, http://www.itu.int/net/home/index.aspx

[Iyengar]   R. Iyengar, K. Kar, B. Sikdar, "Scheduling algorithms for point-to-multipoint operation in IEEE 802.16 networks," Proc. 2nd Workshop on Resource Allocation in Wireless Networks (RAWNET), April 2006

[Jacobson]   V. Jacobson, "Congestion Avoidance and Control", Proc. 1988 SIGCOMM Symposium on Communications Architectures and Protocols, pages 314–329, Stanford, CA, August 1988.

[Jha]   S. Jha, M. Hassan, "Engineering Internet QoS", Artech House, 2002

[JSAC]   IEEE Journal on Selected Areas in Communications, "Traffic Engineering for Multi-layer Networks", Jun. 2007

[Joost]   Joost Inc., http://www.joost.com/

[Keshav]   S. Keshav, "An Engineering Approach to Computer Networking: ATM Networks, the Internet, and the Telephone Network", Addison-Wesley, May 1997

[Kleeman]   M. Kleeman, "Point of Disconnect: Internet Traffic and the U.S. Communications Infrastructure", International Journal of Communication 1, pp. 177–183, Feb. 2007.

[Ma]   Q. Ma, "QoS Routing in the Integrated Services networks", Ph.D. dissertation, CMU-CS-98-138, Jan. 1998.

[Mathis]   M. Mathis, J. Semke, J. Mahdavi and T. Ott, "The Macroscopic Behavior of the TCP Congestion Avoidance Algorithm", Computer Communication Review, vol. 27, No. 3, pp. 67–82, July 1997.

[McastOps]   K. Kompella, "Overcoming the Challenges of Multicast", Carrier Ethernet World Congress (CEWC) presentation, Sept 2007

[MC2001]   "U.S Communications Infrastructure at A Crossroads: Opportunities Amid the Gloom", McKinsey & Company for Goldman-Sachs, August 2001.

[MEF]   Metro Ethernet Forum, http://metroethernetforum.org

[Metanoia]   Metanoia Inc., http://www.metanoia-inc.com/

[MITcfp]   MIT Communications Futures Program, inter-provider support for quality of service, http://cfp.mit.edu/groups/internet/qos.html

[ML2002]   "Optical Systems", Merril Lynch Technical Report, April 2002.

[Mogul]   J. Mogul, F. Douglis, A. Feldmann, and B. Krishnamurthy, "Potential benefits of delta encoding and data compression for HTTP", Proc. SIGCOMM '97 Conference, pages 181–194, Cannes, France, September 1997.

[Mozilla]   Mozilla.org, http://www.mozilla.org/

[MRTG]   Multi-Router Traffic Grapher, http://www.mrtg.com/

[NANOG]   North American Network Operators' Group (NANOG), http://www.nanog.org/

[NCP]   Network Control Program, http://en.wikipedia.org/wiki/Network_Control_Program

[Nessoft]   What's "Normal" for latency and packet loss?, http://www.nessoft.com/kb/42

[Netflix]   Netflix Online Movie Rentals, http://www.netflix.com

[Netflow]   "Cisco IOS Netflow introduction", http://www.cisco.com/en/U.S./products/ps6601/products_ios_protocol_group_home.html

[NNcerf]   Testimony of Mr. Vinton G. Cerf, United States Senate Committee on the Judiciary Hearing on Reconsidering our Communications Laws, June 14, 2006, http://judiciary.senate.gov/testimony.cfm?id=1937&wit_id=5416

[NNeu]   Commission Of the European Communities, "Impact Assessment for The Regulatory Framework For Electronic Communications", http://ec.europa.eu/information_society/policy/ecomm/doc/library/proposals/ia_en.pdf

[NNfcc]   "New Principles Preserve and Promote the Open and Interconnected Nature of Public Internet", http://hraunfoss.fcc.gov/edocs_public/attachmatch/DOC-260435A1.pdf

[NNhahn]   The Economics of Net Neutrality, http://www.aei-brookings.org/admin/authorpdfs/page.php?id=1269

[NNLessig] http://www.lessig.org/blog/archives/Lessig_Testimony_2.pdf

[NNpoll]   "Net Neutrality Rises Again", http://www.fool.com/investing/general/2007/01/18/Net Neutrality-rises-again.aspx

[NNregist] "Google snubs Net Neutrality debate", http://www.theregister.co.uk/2007/03/20/uk_net_neutrality/, The Register, March 2007

[NNwiki]   http://en.wikipedia.org/wiki/Network_neutrality_in_the_U.S.

[NNwu]   T. Wu, "Net Neutrality, Broadband Discrimination", Journal of Telecommunications and High Technology Law, Vol. 2, p. 141, 200, http://www.cdt.org/speech/Net Neutrality/2005wu.pdf

[Odlyzko]   A. Odlyzko, "The economics of the Internet: Utility, utilization, pricing, and Quality of Service", July, 1998. http://www.dtc.umn.edu/~odlyzko/doc/internet.economics.pdf

[Odlyzko1] A. Odlyzko, "Smart and stupid networks: Why the Internet is like Microsoft", ACM Networker, 2 (5), December 1998.

[Odlyzko2] A. Odlyzko, "Data Networks are Lightly Utilized, and will Stay that Way", Review of Network Economics, 2 (no. 3), September 2003, pp. 210–237. http://www.rnejournal.com/articles/andrew_final_sept03.pdf

[Odlyzko3] A. Odlyzko, "The history of communications and its implications for the Internet", AT&T Labs research report, http://www.dtc.umn.edu/~odlyzko/doc/history.communications0.pdf

[Odlyzko4] A. Odlyzko, "Content is Not King", First Monday, Jan. 2001, http://www.firstmonday.org/issues/issue6_2/odlyzko/

[Ofcom]   Office of Communications, "Strategic Review of Telecommunications Phase 2 Proposals", http://www.ofcom.org.uk/media/news/2004/11/nr20041118

[OPNET]   OPNET Inc. http://www.opnet.com/

[OPsec]   Merike Kaeo, "Operational Security Current Practices", draft-ietf-opseccurrent-practices-03, May 2006.

[Park]   K.I. Park, "QoS in Packet Networks"", The International Series in Engineering and Computer Science, 2004

[Patterson] D. Patterson, et al, "Recovery-Oriented Computing (ROC): Motivation, Definition, Techniques, and Case Studies", U.C. Berkeley Computer Science Technical Report, March, 2002.

[PCCW]   http://www.pccw.com/eng/Products/ForYourHome/BroadbandInternetandTVServices.html

[PingER]   http://www-iepm.slac.stanford.edu/pinger/

[PPPoE]   Point-to-Point Protocol over Ethernet, http://en.wikipedia.org/wiki/Point-to-Point_Protocol_over_Ethernet

[P2P]	Peer-to-peer, http://en.wikipedia.org/wiki/Peer-to-peer
[P800]	Methods for subjective determination of transmission quality, ITU-T Recommendation P.800, 1996.
[QBarch]	http://qos.internet2.edu/wg/documents-informational/draft-i2-qbone-arch-1.0/
[QBone]	http://qbone.internet2.edu/
[QBSS]	Shalunov, S., Teitelbaum, B., "QBone Scavenger Service (QBSS) Definition", Internet2 Technical Report, Proposed Service Definition, Internet2 QoS Working Group Document, March, 2001.
[Q.Ni]	Q. Ni, "Performance analysis and enhancements for IEEE 802.11e Wireless Networks," IEEE Network, vol., no. , July/August 2005, pp. 21–27.
[QoS-DoS]	S. Shalunov and B. Teitelbaum, "Quality of Service and Denial of Service", ACM SIGCOMM Revisiting IP QoS Workshop: Why do we care, what have we learned? (RIPQOS), August 2003
[QoSrisk]	B. Teitelbaum and S. Shalunov, "What QoS Research Hasn't Understood About Risk", ACM SIGCOMM Revisiting IP QoS (RIPQOS) Workshop: Why do we care, what have we learned? (RIPQOS), August 2003
[RACS]	Resource and Admission Control and IMS, http://www.itu.int/ITU-T/worksem/ngn/200603/presentations/s4_mainwaring.pdf
[Rath1]	H. Rath, A. Bhorkar and V. Sharma, "An opportunistic uplink scheduling scheme to achieve bandwidth fairness and delay for multi-class traffic in WiMAX (802.16) broadband networks," IEEE Globecom'06, San Francisco, CA, 27 Nov. – 1 Dec. 2006.
[Rath2]	H. Rath, A. Karandikar, and V. Sharma, "Adaptive Modulation-based TCP-Aware Uplink Scheduling in IEEE 802.16 Networks," to appear IEEE ICC'08, May 2008.
[RED]	S. Floyd and V. Jacobson, "Random Early Detection gateways for Congestion Avoidance", IEEE/ACM Transactions on Networking, Vol. 1. No. 4, August 1993, pp. 397–413.
[Reding]	ADSLgr.com, "Interview with Viviane Reding, EU commissioner for Information Society and Media", http://adslgr.com/forum/showthread.php?t=137431
[RFC114]	A. Bhushan, "A File Transfer Protocol", RFC 114, April 1971. (Obsoleted by RFC 959).
[RFC741]	D. Cohen, "Specifications for the Network Voice Protocol (NVP)", RFC 741, Nov. 1977
[RFC791]	J. Postel, "Internet Protocol", STD 5, RFC 791, September 1981.
[RFC793]	J. Postel, "Transmission Control Protocol", STD 7, RFC 793, Sept. 1981.
[RFC795]	J. Postel, "Service Mappings", RFC 795, September 1981.
[RFC1195]	R. Callon, "Use of OSI IS-IS for Routing in TCP/IP and Dual Environments", RFC 1195, Dec. 1990.
[RFC1633]	R. Braden, D. Clark, and S. Shenker, "Integrated Services in the Internet Architecture: an Overview", RFC 1633, June 1994.
[RFC1866]	T. Berners-Lee, D. Connolly, "Hypertext Markup Language – 2.0", RFC 1866, Nov. 1995.
[RFC1925]	R. Callon, "The Twelve Networking Truths", RFC 1925, April 1996.
[RFC1958]	B. Carpenter, "Architectural principles of the Internet", RFC 1958, June 1996.
[RFC2205]	R. Braden, L. Zhang, et al, "Resource ReSerVation Protocol (RSVP) -Version 1 Functional Specification", RFC 2205, Sept. 1997.

[RFC2211]   J. Wroclawski, "Specification of the Controlled-Load Network Element Service", RFC 2211, Sept. 1997

[RFC2212]   S. Shenker, C. Partridge and R. Guerin, "Specification of Guaranteed Quality of Service", RFC 2212, Sept. 1997

[RFC2309]   B. Braden et al., "Recommendation on Queue Management and Congestion Avoidance in the Internet", RFC 2309, Apr. 1998

[RFC2328]   J. Moy, "OSPF Version 2", RFC 2328, Apr. 1998.

[RFC2386]   E. Crawley, R. Nair, B. Jajagopalan and H. Sandick, "A Framework for QoS-based Routing in the Internet", RFC 2386, Aug. 1998.

[RFC2460]   Deering, S. and R. Hinden, "Internet Protocol, Version 6 (IPv6) Specification", RFC 2460, December 1998.

[RFC2474]   Nichols, K., Blake, S., Baker, F. and D. Black, "Definition of the Differentiated Services Field (DS Field) in the IPv4 and IPv6 Headers", RFC 2474, December 1998.

[RFC2475]   Black, D., Blake, S., Carlson, M., Davies, E., Wang, Z. and W. Weiss, "An Architecture for Differentiated Services", RFC 2475, December 1998.

[RFC2547]   E. Rosen and Y. Rekhter, "BGP/MPLS VPN", RFC 2547, March 1999.

[RFC2581]   M. Allman, V. Paxson, W. Stevens, "TCP Congestion Control", RFC 2581, April 1999

[RFC2597]   Heinanen, J., Baker, F., Weiss, W. and J. Wroclawski, "Assured Forwarding PHB Group", RFC 2597, June 1999.

[RFC2598]   V. Jacobson, K. Nichols, and K. Poduri, "An Expedited Forwarding PHB", RFC 2598, June 1999.

[RFC2616]   R. Fielding, J. Gettys, J. Mogul, H. Frystyk, L. Masinter, P. Leach and T. Berners-Lee, "Hypertext Transfer Protocol – HTTP/1.1", RFC2616, Jun. 1999.

[RFC2679]   G. Almes, S. Kalidindi and M. Zekauskas, "A One-way Delay Metric for IPPM", RFC 2679, Sept. 1999.

[RFC2680]   G. Almes, S. Kalindi and M. Zekauskas, "A One-way Packet Loss Metric for IPPM", RFC 2680, Sept. 1999.

[RFC2702]   D. Awduche, J. Malcolm, J. Agogbua, M. O'Dell and J. McManus, "Requirements for Traffic Engineering Over MPLS", RFC 2702, Sept. 1999

[RFC2814]   R. Yavatkar et al. "Subnet Bandwidth Manager (SBM): A Protocol for RSVP-based Admission Control", RFC2814, May 2000.

[RFC2815]   M. Seaman, A. Smith, and E. Crawley, "Integrated Service Mappings on IEEE 802 Networks", RFC 2815, May 2000.

[RFC2873]   X. Xiao, A. Hannan, V. Paxson and E. Crabbe, "TCP Processing of the IPv4 Precedence Field", RFC 2873, June 2000.

[RFC2998]   Y. Bernet, "A Framework for Internet Services Operation over DiffServ Networks", RFC 2998, November 2000.

[RFC3031]   E. Rosen, A. Viswanathan and R. Callon, "Multiprotocol Label Switching Architecture", RFC 3031, Jan. 2001.

[RFC3209]   D. Awduche, et al, "RSVP-TE: Extensions to RSVP for LSP Tunnels", RFC 3209, Dec. 2001

[RFC3246]   B. Davie, A. Charny, et al, "An Expedited Forwarding PHB (Per-Hop Behavior)", RFC 3246, March 2002.

[RFC3248]   G. Armitage, B. Carpenter, et al, "A Delay Bound Alternative Revision of RFC 2598", RFC 3248, March 2002.

[RFC3270]   F. Le Faucheur, et al, "Multi-Protocol Label Switching (MPLS) Support of Differentiated Services", RFC 3270, May 2002

[RFC3272]   D. Awduche, A. Chiu, A. Elwalid, I. Widjaja and X. Xiao, "Overview and Principles of Internet Traffic Engineering", RFC 3272, May 2002

[RFC3393]   C. Demichelis and P. Chimento, "IP Packet Delay Variation Metric for IP Performance Metrics (IPPM)", RFC 3393, Nov. 2002

[RFC3564]   F. Le Faucheur and W. Lai, "Requirements for Support of Differentiated Services-aware MPLS Traffic Engineering", RFC 3564, Jul. 2003.

[RFC3439]   R. Bush and D. Meyer, "Some Internet Architectural Guidelines and Philosophy", Dec. 2002

[RFC4090]   P. Pan, G. Swallow and A. Atlas, "Fast Reroute Extensions to RSVP-TE for LSP Tunnels", RFC4090, May 2005

[RFC4271]   Y. Rekhter, T. Li and S. Hares, "A Border Gateway Protocol 4 (BGP-4)", January 2006

[RoutArch]   S. Halabi, D. McPherson, "Internet Routing Architectures (2nd Edition)", Cisco Press, 2000.

[RPR]   http://en.wikipedia.org/wiki/Resilient_Packet_Ring

[Saltzer]   End-To-End Arguments in System Design, J.H. Saltzer, D.P.Reed, D.D.Clark, ACM TOCS, Vol 2, Number 4, November 1984, pp. 277–288.

[SaveNet]   Save the Internet, http://www.savetheinternet.com/

[Savvis]   http://www.savvis.net/corp/Products+Services/Digital+Content+Services/Content+Delivery+Services/

[Scott]   D. Scott, "Making Smart Investments to Reduce Unplanned Downtime", Tactical Guidelines, TG-07-4033, Gartner Group Research Note, March 1999.

[SecEfforts]   C. Lonvick and D. Spak, "Security Best Practices Efforts and Documents", draft-ietf-opsec-efforts-03.txt, April 2006.

[Sharma1]   V. Sharma and N. Vamaney, "The uniformly-fair deficit round-robin scheduler for IEEE 802.16 WiMAX networks," IEEE Milcom'07, Orlando, FL, 29–31 October 2007.

[Sharma2]   V. Sharma and A. Karandikar, "Elements of Cross-Layer System and Network Design for QoS-Enabled WiMAX Networks" Tutorial T5, IEEE Milcom'07, Orlando, FL, 29–31 October 2007.

[Soldani]   D. Soldani, "QoS and QoE Management in UMTS Mobile Networks: An Introduction to Service Planning, Provisioning, Performance Monitoring and Optimisation", John Wiley & Sons, 2006

[Szigeti]   T. Szigeti and C. Hattingh, "End-To-End QoS Network Design: Quality of Service in LANs, WANs, and VPNs", Cisco Press, 2004

[TISPAN]   Telecoms & Internet converged Services & Protocols for Advanced Networks, http://www.etsi.org/tispan/

[TraMatr]   T. Telkamp, "Best Practices for Determining the Traffic Matrix", NANOG 35, Oct. 2005, http://www.cariden.com/technologies/papers/nanog35-tm.pdf

[UPS]   United Parcel Service of America, Inc., http://www.ups.com/

[Vegesna]   S. Vegesna, "IP Quality of Service", Cisco Press, 2001.

[Verisign]   Verisign Inc. http://www.verisign.com/

[VoIPinfo]   http://www.voip-info.org/wiki/view/QoS

[VonageF]   "S. Korea to block some Internet calling", http://www.vonage-forum.com/ftopic15054.html

[VZeth]   Verizon Business Ethernet Services, http://www.verizonbusiness.com/us/data/ethernet/compare.xml

[WANDL]   Wide Area Network Design Laboratory (WANDL), http://www.wandl.com/

[Wang]   Z. Wang, "Quality of Service in IP Networks", Elsevier/Morgan Kaufmann, 2001.

[Wang1]   Z. Wang, "Routing and Congestion Control in Datagram Networks", Ph.D. dissertation, Dept. of Computer Sci., University College London, Jan. 1992.

[Wang2]   Z. Wang, "Internet QoS: Architectures and Mechanisms for Quality of Service", Morgan Kaufmann, 2001.

[Wang3]   Y. Wang, Z. Wang and L. Zhang, "Internet traffic engineering without full mesh overlaying", INFOCOM 2001, April 2001

[Willinger]   W. Willinger and J. Doyle, "Robustness and the Internet: Design and evolution", 2002. http://netlab.caltech.edu/pub/papers/part1_vers4.pdf

[WRED]   Weighted random early detection, http://en.wikipedia.org/wiki/Weighted_random_early_detection

[WSHpost1]   Washington Post, "Executive Wants to Charge for Web Speed", Dec. 1, 2005, http://www.washingtonpost.com/wp-dyn/content/article/2005/11/30/AR2005113002109.html

[WSHpost2]   Washington Post, "Verizon's Executive Calls for End to Google's 'Free Lunch'", Feb 7, 2006, http://www.washingtonpost.com/wp-dyn/content/article/2006/02/06/AR2006020601624.html

[WSHpost3]   Washington Post, "SBC Head Ignites Access Debate", http://www.washingtonpost.com/wp-dyn/content/article/2005/11/03/AR2005110302211.html

[Xiao1]   X. Xiao, L.M. Ni, "Internet QoS: A Big Picture", IEEE Network, Jan. 1999.

[Xiao2]   X. Xiao, A. Hannan, B. Bailey and L.M. Ni, "Traffic engineering with MPLS in the Internet", IEEE Network, Mar. 2000

[Xiao3]   X. Xiao, "Providing Quality of Service in the Internet", Ph.D. Dissertation, Michigan State University, May 2000.

[Y.Xiao]   Y. Xiao, "IEEE 802.11e: QoS Provisioning at the MAC Layer", IEEE Wireless Communications, June 2004, pp. 72-79.

[2Bit]   K. Nichols, V. Jacobson and L. Zhang, "A Two-bit Differentiated Services Architecture for the Internet", ftp://ftp.ee.lbl.gov/papers/dsarch.pdf, November, 1997.

[802.1D]   "MAC Bridges", ISO/IEC 10038, ANSI/IEEE Std 802.1D-1993. http://www.ieee802.org/1/pages/802.1D.html

[802.11]   IEEE Wireless LAN standards, http://en.wikipedia.org/wiki/IEEE_802.11

[802.11abgn]   IEEE Wireless LAN standards, http://en.wikipedia.org/wiki/IEEE_802.11

[802.11e]   "Quality of Service Enhancements for Wireless LAN", IEEE 802.11e, 2005, http://en.wikipedia.org/wiki/802.11e

[802.16]   IEEE Broadband Wireless Access Standards, September 2004, http://en.wikipedia.org/wiki/802.16

Index

275

Printed and bound by CPI Group (UK) Ltd, Croydon, CR0 4YY

03/10/2024

01040320-0004